Gas-Adsorption Chromatography

Gas-Adsorption Chromatography

Andrei V. Kiselev and Yakov I. Yashin

Department of Chemistry, Moscow University
and
Institute of Physical Chemistry, Academy of Sciences of the USSR

Translated from Russian by
J. E. S. Bradley
Senior Lecturer in Physics, University of London

℗ Springer Science+Business Media, LLC 1969

Professor Andrei Vladimirovich Kiselev is the director of the Laboratory of Adsorption and Gas Chromatography, Department of Chemistry, Moscow University, and also of the Laboratory of Surface Chemistry, Institute of Physical Chemistry, Academy of Sciences of the USSR. He was born in 1908; in 1938 he attained the degree of candidate and in 1950 the degree of doctor of chemical sciences. He has published over 500 papers on the surface chemistry of solids, the synthesis and structure of adsorbents, the theory of molecular interaction in adsorption and chromatography, thermodynamics of adsorption, calorimetric measurements of heats of adsorption, and adsorption from solution.

Yakov Ivanovich Yashin is the director of the gas chromatography division of an experimental automatic equipment design organization. Born in 1936, he graduated from the chemical faculty at Gor'kii University in 1958 and attained the degree of candidate of chemical sciences in 1965. Since 1959 he has worked in the area of gas chromatography, since 1961 in collaboration with Professor Kiselev. He has published over 50 papers on the geometrical structure and chemical nature of adsorbent surfaces, on molecular interactions as studied by gas chromatography, on applications of gas chromatography in analysis and in physical chemistry, and on the design of laboratory chromatographs.

Library of Congress Catalog Card Number 69-12531

The original Russian text, published for the Institute of Physical Chemistry of the Academy of Sciences of the USSR by Nauka Press in Moscow in 1967 as part of a series on the Chemistry of Surfaces and Adsorption, has been corrected by the authors for the English edition.

Андрей Владимирович К и с е л е в, Яков Иванович Я ш и н
Газо-адсорбционная хроматография
GAZO-ADSORBTSIONNAYA KHROMATOGRAFIYA
GAS-ADSORPTION CHROMATOGRAPHY

© 1969 Springer Science+Business Media New York

Originally published by Plenum Press, New York in 1969.

United Kingdom edition published by Plenum Press, London
A Division of Plenum Publishing Company, Ltd.
Donington House, 30 Norfolk Street, London W. C. 2, England

ISBN 978-1-4899-6238-6   ISBN 978-1-4899-6503-5 (eBook)
DOI 10.1007/978-1-4899-6503-5
Softcover reprint of the hardcover 1st edition 1969

Foreword

One of the present authors in collaboration with his colleagues has undertaken the compilation of a series of books on adsorption and the chemistry of solid surfaces, because this area has recently developed so rapidly that it is no longer possible to deal with it in detail in one volume. This is the more important because this area has become important in chemical technology and in many aspects of research in chemistry, biochemistry, and elsewhere. The object in this series is not to give an exhaustive exposition of all lines of work or to survey all publications; the books will rather deal mainly with those areas in surface chemistry and adsorption in which the authors have direct practical experience. The present volume deals with gas–adsorption chromatography.

Molecular chromatography, especially gas chromatography, would appear one of the most promising areas, since gas chromatography has become one of the principal instrumental methods of analysis in many branches of industry and research. The simple physicochemical basis (differences in molecular interactions of the components in solution or adsorption) makes the method universal; it is also highly effective for complex mixtures of organic and inorganic substances of boiling point up to 600°C. Rapid automatic analysis is possible, with ionization detectors of high sensitivity, while the apparatus is simple and may be standardized. Automation for many chemical processes is often based on gas chromatography, as well as routine plant analyses.

The rapid development of gas chromatography followed the discovery by Martin and James of gas–liquid chromatography 15 years ago, and the pace of development is continually quickening. The current annual output of original papers on gas chromatography is in excess of 2000. The method is now often applied to mixtures that could not be analyzed by other methods.

Development gas-adsorption and gas−liquid forms of the method are the main ones used in analysis; but the usual highly active adsorbents of high specific surface are suitable only for separating highly volatile substances, whereas far less volatile substances are separated in gas−liquid chromatography. The latter is therefore the more widely used.

Most books on gas chromatography deal mainly with gas−liquid chromatography, as in Keuleman's (1959) Gas Chromatography, Zhukhovitskii and Turkel'taub's (1962) Gas Chromatography, and Nogare and Juvet's (1966) Gas−Liquid Chromatography.

However, various deficiences in this method have become apparent with the increase in sensitivity of detectors and with the extension of applications, in particular the volatility and instability of liquid phases at high column temperatures. Recently there have been advances in the synthesis and modification of solid adsorbents with uniform surfaces having a variety of compositions, which has increased the interest in gas-adsorption chromatography. Unfortunately, the papers on this are scattered over a variety of journals and are often inaccessible. One objective of this book is to overcome this difficulty.

Gas chromatography is still in the stage of empirical selection, so we have considered it important here to deal with gas-adsorption chromatography from a unified standpoint based on molecular interactions in adsorption.

It is assumed that the reader is familiar with the general theory and apparatus of gas chromatography; these are adequately dealt with in the books mentioned above, and also in Schay's (1963) Theoretical Principles of Gas Chromatography.

The first chapter of this book deals with the gas-adsorption and gas−liquid forms, with emphasis on their advantages and disadvantages, ways of overcoming the disadvantages, ways of utilizing the methods, and areas of application. The second chapter uses the theory of molecular interaction in adsorption to consider the effects of surface chemistry and of molecular structure on gas-chromatographic separation. In that chapter a classification of compounds and sorbents is given on the basis of these features, with a statement of the best uses of nonspecific and specific adsorbents of various types. The third chapter deals in detail with the effects of surface geometry, porosity, and granularity on the separation. The fourth chapter deals with the use of gas-adsorption chromatography in physicochemical studies of adsorption and in the determination of specific surface. The fifth chapter surveys in detail the analytical applications, starting with the separation of hydrogen isomers and isotopes and extending to the analysis of heavy polynuclear aromatic compounds with boiling points up to 600°C. One section of this chapter compares gas-adsorption and gas-liquid methods on the same mixtures. The sixth chapter gives practical techniques for the preparation of adsorbents and adsorption columns.

The book does not deal with the dynamics of gas-adsorption columns. Rachinskii (1964) in Introduction to the General Theory of Sorption Dynamics in Chromatography gives a general treatment of this; Giddings (Dynamics of Chromatography, Vol. 1, Marcel Dekker, Inc., New York, 1965) deals with it in detail. Timofeev (1962) deals with the aspects of adsorption kinetics important to gas-adsorption chromatography in his Adsorption Kinetics.

There seems little doubt that the next few years will see a rapid development of the molecular chromatography of liquid solutions without conversion to vapor, with the use of high-sensitivity methods of detection. This is essential to the chromatographic analysis of complicated and thermally unstable molecules, especially macromolecules, since these are of especial importance to biochemistry and polymer chemistry. Here molecular adsorption from solution onto solid adsorbents will play an important part. However, this topic falls outside the scope of this book, and it is hoped to deal with it separately.

The book is intended for general use in plant laboratories and research institutes, as well as for specialists in adsorption, molecular chromatography, and molecular interactions. It is also of value to graduate students and those specializing in areas of adsorption, chromatography, and analytical, preparative, and industrial applications of gas chromatography

The listing of literature on gas-adsorption chromatography is fairly exhaustive (up to 1966); but this is a difficult task in this rapidly developing field, so it may well be that some papers have been overlooked. We shall be glad to receive notice of such papers and of any other deficiencies in the book.

In the English edition the authors have made several additions to the text which partly cover the recent literature.

We wish to record our thanks to A. P. Arkhipova, A. G. Bezus, T. I. Goryunova, T. S. Kiseleva, R. S. Petrova, and V. L. Yashina for assistance in preparing the manuscript for publication.

A. V. Kiselev and Ya. I. Yashin

Contents

Chapter I

Introduction. Advantages and Disadvantages of Gas-Adsorption Chromatography and Gas—Liquid Chromatography

Tsvet [2] founded a chromatography in adsorption form in 1903. Some topics closely related to frontal gas-adsorption chromatography were considered without relation to gas chromatography in connection with the performance of gas masks [3] (see also [4, 5]), but gas-chromatographic analysis of mixtures began to develop rapidly only from 1952, when James and Martin [6] described the development form of gas—liquid partition chromatography. This has since been the main method used in analysis. Gas—liquid partition has advantages over gas adsorption in that a wide range of fixed liquids can be used and also that a liquid is homogeneous, so the solubility isotherms are virtually linear over a wide range in concentration starting with the very lowest. The range of commercially available adsorbents is very small; moreover, these adsorbents are geometrically and chemically inhomogeneous.

However, some disadvantages of gas—liquid partition become apparent, especially with the increase in sensitivity of detectors, extension of the operating temperature range, and application of gas chromatography for automatic monitoring and detection of impurities. These concern especially the volatility and instability of the liquid phase [7, 8], which make it impossible to operate at high temperatures [9], interfere with the detection of trace components [10], and prevent the use of gas—liquid columns for prolonged automatic analysis [11], for preparative purposes [13], and for methods involving temperature programming [12].

1

Gas adsorption is free from these disadvantages. Its main disadvantage is nonlinearity in the adsorption isotherms, which causes asymmetry in peaks; this is due to the geometrical and chemical inhomogeneity of the usual active adsorbents, and it is especially pronounced for strongly adsorbed substances. This inhomogeneity of the usual active adsorbents is accompanied by strong adsorption and often also by considerable catalytic activity; these features restrict the use of the usual adsorbents, which therefore are used mainly with gaseous materials having no active functional groups, since these give nearly linear isotherms at the concentrations and temperatures commonly employed. Papers appearing in 1947-1954 (especially those of Claesson [14], Phillips [15], Turkel'-taub [16], Cremer [17], Janak [18], and Ray [19]) left the impression, which persisted up till about 1960, that gas adsorption was a supplement to gas–liquid partition for separating gases and low-boiling liquids, where the separation produced by liquids is inadequate on account of low solubility [20].

Applications to automatic process control accentuated the demand for speed of analysis [11]; often the time required is only 0.5-5 min. In this respect gas–liquid partition has no advantages over gas adsorption, because the separation produced by a fixed phase (liquid or solid) is dependent on the surface area, selectivity, and exchange rate (rates of solution and evaporation, or of adsorption and desorption). If the partition isotherms near the working concentration are linear, the separation is the best when the mass-transfer coefficient is large. Exchange is the main cause of peak broadening in the gas–liquid method if the flow rate of the carrier gas is high, and the principal rate-limiting factors here are passage through the interface [21, 22] and diffusion within the liquid film [23]. The time for molecular adsorption is usually very small [24], so the adsorption rate is governed mainly by diffusion in the gas to the adsorbent: to the outer surface for a nonporous material and to the inner surfaces in the porous one. Desorption is rather slower, but the desorption time is also short at high temperatures, especially for nonporous and fairly macroporous materials. This means that a nonporous or macroporous adsorbent allows high linear speeds of the carrier gas to be used without loss of performance [25]. The performance of a column filled with such an adsorbent is only slightly dependent on the speed of the carrier gas, so programmed gas flow can be used in high-speed analysis [86].

Inhomogeneity in the immobile phase does occur in gas–liquid chromatography. It is very difficult to produce a uniform liquid film on the usual inhomogeneously porous carriers, and so it is difficult to avoid broadening due to variations in film thickness [26]. Moreover, adsorption and catalysis by the surface of the carrier sometimes play a part [27, 28].

The molecular geometry greatly influences the adsorption energy, so selective columns can be based not only on differences in the electronic structures of the components but also on differences in molecular geometry during the adsorption on the fairly smooth surface of the adsorbent [29, 30].

This means that adsorption gives considerable scope for rapid analysis with columns of high performance [31]. Here it is necessary to minimize the additional broadening due to variation in the adsorption lifetime from one part of the surface to another, i.e., to chemical and geometrical inhomogeneity. This is comparatively easy with nonporous or macroporous adsorbents; suitable choice of working temperature and column length can also assist. In gas–liquid partition we get adsorption at the surface of the liquid as well as solubility in the liquid, and the former may make a considerable contribution to the retention time, especially if the liquid is highly polar [32-34]. This applies even when the liquid is a macromolecular compound of high viscosity, when diffusion within the liquid is slow. Adsorption may thus play a part comparable with that of solubility.

Difficulties have been encountered in developing and applying the theory of gas-adsorption chromatography [35], which were greater than those for gas–liquid chromatography partly because there is no reliable evidence on the relation of adsorption to chemical or geometrical structure, and partly also because of the lack of reproducibility between batches and the limited range of surfaces available. These difficulties are not now insuperable [36], whereas the above disadvantages of the gas–liquid method go to the root of the technique. Kiselev [37-41] has pointed out the need to improve adsorbents in this connection.

Many methods have been described [39-43] for producing symmetrical peaks by linearizing the adsorption isotherms. The most important of these are as follows:

1. Preparation of crystalline nonporous finely divided adsorbents by heat treatment (graphitized carbons [36, 44, 45]), distillation (NaCl [46]), and precipitation (various salts [47]).

2. Use of crystalline porous adsorbents, e.g., zeolites (see [48] for review) and other porous salts [49, 50], and also porous polymers [87].

3. Preparation of macroporous adsorbents with uniform large pores by controlled synthesis or by geometrical modification of the usual active adsorbents [38, 39, 51-57].

4. Chemical modification of the inhomogeneous surface of a large-pore adsorbent by treatment with surface-active compounds having the appropriate properties [37, 58, 59],

5. Production of adsorption centers of a single type by deposition on a large-pore adsorbent of solid organic [60] or inorganic substances, e.g., palladium black on cellite (for specific adsorption of hydrogen [61]), graphitized carbon on modified weakly adsorbing silica gel [62], on porous Teflon [88], alkalis [63, 64], inorganic salts [65-67], and eutectic salt mixtures [68, 69] on various carriers, especially for analyses at high temperatures [70].

6. Performance of analyses at relatively high temperatures (100°C or more above the boiling point) [71], since this tends to linearize the isotherms even for an inhomogeneous surface [72].

7. Deposition of monomolecular layers of strongly adsorbed substances, especially polymers [59, 73], on the strongly adsorbing surfaces of a nonporous of macroporous adsorbent. This means that adsorption on the solid is replaced by adsorption on the monolayer, whose volatility is much reduced by the adsorption forces. It is very effective to deposit crystallizable organic compounds on thermal carbon black: benzophenone [89], anthraquinone [90], and the nonvolatile phthalocyanins of Ca, Co, Zn, and Ni [91-93]. In the last case it is possible to perform many separations that cannot be performed, for example, by gas – liquid chromatography.

8. Blocking of the most active parts of an inhomogeneous surface with a liquid [74, 75]; and

9. Blocking of the active centers of an inhomogeneous surface with an adsorbed carrier gas (e.g., NH_3 or CO_2) [76].

10. Use of gases at pressures up to 50 atm, or substances in the supercritical state, as mobile phases in the analysis of heavy molecules [94, 95].

Of these nine methods, 1-4 (sometimes with 5) and 6 and 7 are the most promising. The fifth method (deposition of more homogeneous solid adsorbents) may be considered as a combination of 3 and 4, since two objects are met: production of uniform pores (geometrical modification) and production of an adsorbing surface of a single type (chemical modification). Method 7 eliminates the variation in adsorption energy and provides a simple and convenient way of adjusting the chemical nature of the surface (that of the monolayer). In this case the gas-adsorption method uses an advantage of the gas-liquid method (variety of deposited compounds) while largely overcoming the disadvantage of volatility, especially with heated columns, since the monolayer is strongly adsorbed on the carrier and thus has its volatility greatly reduced.

Raising the temperature is particularly beneficial when the solid adsorbent has been suitably chosen as to nature, specific surface, and pore size [77].

Blocking with small quantities of liquid provides a method lying somewhere between gas adsorption, adsorption on a monolayer (method 7), and gas-liquid partition. Method 9 (strongly adsorbed carrier gas) is not always convenient, and sometimes is quite impracticable, e.g., when high-sensitivity ionization detectors are used. The last method is promising for the separation and analysis of compounds of molecular weight over 1300. The method is intermediate between classical gas and liquid methods, since the mobile phase is appreciably adsorbed.

Gas adsorption can be used with high-performance capillary columns (internal surface porous). A strongly adsorbing porous layer may be produced either from the capillary material itself (e.g., borosilicate glass) [78-80] or by deposition of particles of a suspension [81], a sol [82], or a gel [83]. For example, a glass column with a porous surface layer produced by corrosion of the glass proved capable of completely separating the isotopes and isomers of hydrogen [79] and a mixture of all the deuteromethanes [80]. Capillary columns filled with active adsorbents have also been used [84, 85]. Glass capillary columns have been made with thick coatings of thermal black on the walls and a free lumen; these have been used to separate various ordinary and deuterated compounds [96].

Gas-adsorption chromatography thus has considerable scope for development by the use of crystalline materials (porous and nonporous) homogeneous macroporous amorphous materials, and modified solid adsorbents. The last section of the book deals with a comparison of gas-adsorption and gas-liquid methods for certain liquid mixtures.

LITERATURE CITED

1.   C. G. Scott and C. S. G. Phillips, in Collection: Gas Chromatography 1964, A. Goldup (ed.), London, 1965, p. 129.
2.   M. S. Tsvet, Trudy Varshav. Obshch. Estest., Otdel. Biol., 14:20 (1903).
3.   N. A. Shilov, L. K. Lepin', and S. A. Voznesenskii, Zh. Russk. Fiz.-Khim. Obshch., 61:1107 (1929).
4.   M. M. Dubinin, The Physicochemical Principles of Sorption Technology, Moscow, ONTI, 1935.
5.   M. M. Dubinin and K. V. Chmutov, The Physicochemical Principles of Gas Defense, Moscow, 1939.
6.   A. T. James and A. J. P. Martin, Biochem. J., 50:679 (1952).
7.   S. Dal Nogare and Z. W. Safranski, Anal. Chem., 30:894 (1958).
8.   R. A. Keller, R. Bate, B. Costa, and F. Perry, J. Chromatog., 8:157 (1962).
9.   N. A. Malofeev, I. P. Yudina, and N. M. Zhavoronkov, Usp. Khim., 31:710 (1962).
10.   C. J. Kuley, Anal. Chem., 35:1472 (1963).
11.   H. J. Maier, IRE Trans. Industr. Electron., 9:11 (1962).
12.   L. S. Ettre, R. D. Condon, F. J. Kabot, and E. W. Cieplinski, J. Chromatog., 13:305 (1963).
13.   V. Yu. Zel'venskii, S. A. Volkov, and K. I. Sakodynskii, Neftekhimiya, 7(1):123 (1967).
14.   S. Claesson, Arkiv Kemi, 23:1 (1947).
15.   C. S. G. Phillips, Discussions Faraday Soc., 7:241 (1950).
16.   N. M. Turkel'taub, Zh. Anal. Khim., 5:200 (1950).
17.   E. Cremer and R. Müller, Z. Electrochem., 55:217 (1951).
18.   J. Janak, Chem. Listy, 47:464, 817, 828, 837, 1184, 1348 (1953).
19.   N. H. Ray, J. Appl. Chem., 4:82 (1954).
20.   P. S, Porter and J. F. Johnson, Anal. Chem., 33:1152 (1961).
21.   M. A. Khan, Nature, 186:800 (1960).
22.   M. A. Khan, Lab. Practice, 11:26 (1962).
23.   A. Keulemans, Gas Chromatography [Russian translation], Moscow, IL, 1959, p. 191 [English edition: Second, New York, Reinhold, 1959].
24.   J. H. de Boer, The Dynamical Character of Adsorption [Russian translation], Moscow, IL, 1962, p. 49 [English edition: Oxford University Press, 1953].
25.   A. V. Kiselev and Ya. I. Yashin, in collection: Gas Chromatography, Trans-

actions of the Third All-Union Conference on Gas Chromatography, Dzerzhinsk, Izd. Dzerzhinsk. Fil. OKBA, 1966, p. 131.

26. J. C. Giddings, Anal. Chem., 34:458 (1962).

27. M. Vilkas and N. A. Abraham, Bull. Soc. Chim. France, 1959, p. 1651.

28. D. M. Ottenstein, in collection: Advances in Chromatography, J. C.
Giddings and R. A. Keller (eds.), New York, Marcel Dekker, Inc., Vol. 3,
1966, p. 137.

29. A. V. Kiselev and Ya. I. Yashin, Zh. Fiz. Khim., 40:429 (1966).

30. A. V. Kiselev, G. M. Petov, and K. D. Shcherbakova, Zh. Fiz. Khim., 41(6)
(1967); G. M. Petov and K. D. Shcherbakova, in collection: Gas chromatography 1966, A. B. Littlewood (ed.), London, Institute of Petroleum, 1967,
p. 50.

31. J. C. Giddings, Anal. Chem., 36:1170 (1964).

32. R. L. Martin, Anal. Chem., 33:347 (1961).

33. R. L. Martin, Anal. Chem., 35:116 (1963).

34. R. L. Pecsok, A. de Yllana, and A. Abdul-Karim, Anal. Chem., 36:452 (1964).

35. Physical and Chemical Methods of Separation, E. W. Berg (ed.), New York,
McGraw-Hill, 1963, p. 80.

36. A. V. Kiselev, in collection: Gas Chromatography, Transactions of the
First All-Union Conference on Gas Chromatography, Moscow, Izd. Akad.
Nauk SSSR, 1960, p. 45.

37. A. V. Kiselev, Vestn. Mosk. Gos. Univ., Ser. Khim., 1961, p. 29.

38. A. V. Kiselev, in collection: Gas Chromatography 1962, M. van Swaay (ed.),
London, 1962, p. XXXIV.

39. A. V. Kiselev, in collection: Gas Chromatography, Transactions of the Second All-Union Conference on Gas Chromatography, 1962, Moscow, Izd. Akad.
Nauk SSSR, 1964, p. 31.

40. A. V. Kiselev, in collection: Gas Chromatography, 1964, A. Goldup (ed.),
London, 1965, p. 238.

41. A. V. Kiselev, in collection: Gas Chromatography, Transactions of the Third
All-Union Conference on Gas Chromatography, Dzerzhinsk, Izd. Dzerzhinsk.
Fil. OKBA, 1966, p. 15; in collection: Advances in Chromatography, J.C. Giddings
and R. A. Keller (eds.), New York, Marcel Dekker, Inc., Vol. 4, 1967, p. 113.

42. J. Janak, Ann. New York Acad. Sci., 72:606 (1959).

43. Ya. I. Yaschin, S. P. Zhdanov, and A. V. Kiselev, in collection: Gas Ghro-
matographie 1963, Proceedings of the Fourth Symposium on Gas Chromatography in Leuna, H. P. Angele and H. G. Struppe (eds.), Berlin, Akademie
Verlag, 1963, p. 402.

44. W. D. Schaeffer, W. R. Smith, and M. H. Polley, J. Phys., Chem., 57:469
(1953).

45. A. V. Kiselev, E. A. Paskonova, R. S.Petrova, and K. D. Shcherbako va, Zh.
Fiz. Khim., 38:161 (1964).

46. V. D. Kuznetsov, Surface Energies of Solids, Moscow, Gostekhteorizdat, 1954.

47. J. A. Favre and L. R. Kallenbach, Anal. Chem., 36:63 (1964).

48. A. V. Kiselev, Yu. L. Chernenkova, and Ya. I. Yashin, Neftekhimiya, 5:141
(1965).

49.   A. G. Altenau and L. V. Rogers, Anal. Chem., 36:1726 (1964).

50.   A. G. Altenau and L. V. Rogers, Anal. Chem., 37:1432 (1965).

51.   N. V. Akshinskaya, V. F. Beznogova, A. V. Kiselev, and Yu. S. Nikitin, Zh. Fiz. Khim., 36:2277 (1962).

52.   N. V. Akshinskaya, A. V. Kiselev, and Yu. S. Nikitin, Zh. Fiz. Khim., 37: 927 (1963).

53.   N. V. Akshinskaya, V. Ya. Davydov, A. V. Kiselev, and Yu. S. Nikitin, Kolloidn. Zh., 28:3 (1966).

54.   N. V. Akshinskaya, A. V. Kiselev, E. P. Kolendo, B. A. Lipkind, Yu. S. Nikitin, and Ya. I. Yashin, in collection: Gas Chromatography, Transactions of the Third All-Union Conference on Gas Chromatography, Dzerzhinsk, Izd. Dzerzhinsk. Fil. OKBA, 1966, p. 222.

55.   A. V. Kiselev, Yu. S. Nikitin, R. S. Petrova, K. D. Shcherbakova, and Ya. I. Yashin, Anal. Chem., 36:1526 (1964).

56.   A. V. Kiselev and Ya. I. Yashin, Neftekhimiya, 4:634 (1964).

57.   A. V. Kiselev, I. A. Migunova, V. L. Khudyakov, and Ya. I. Yashin, Zh. Fiz. Khim., 40:2910 (1966).

58.   I. V. Borisenko, A. V. Kiselev, R. S. Petrova, V. K. Chuikina, and K. D. Shcherbakova, Zh. Fiz. Khim., 39:2685 (1965); A. V. Kiselev, Discussions Faraday Soc., 40:205 (1965).

59.   M. Taramasso, Gas Chromatografia, Milano, F. Angeli (ed.), 1966.

60.   C. G. Scott, in collection: Gas Chromatography 1962, M. van Swaay (ed.), London, 1962, p. 36.

61.   E. Glueckauf and G. P. Kitt, in collection: Vapour Phase Chromatography, D. H. Desty (ed.), London, 1956, p. 422.

62.   V. S. Vasil'eva, A. V. Kiselev, Yu. S. Nikitin, R. S. Petrova, and K. D. Shcherbakova, Zh. Fiz. Khim., 35:1386 (1961).

63.   D. A. Vyakhirev, A. I. Bruk, M. V. Zueva, and G. Ya. Mal'kova, Papers on Chemistry and Chemical Technology, Gor'kii, Issue 2, 1964, p. 268.

64.   A. I. Bruk, D. A. Vyakhirev, A. V. Kiselev, Yu. S. Nikitin, and N. M. Olefirenko, Neftekhimiya, 7:145 (1967).

65.   C. G. Scott, in collection: Gas Chromatography 1962, M. van Swaay (ed.), London, 1962, p. 46.

66.   C. G. Scott, J. Gas Chromatog., 4:4 (1966).

67.   O. T. Chortyk, W. S. Schlotzhauer, and R. L. Stedman, J. Gas. Chromatog., 3:394 (1965).

68.   W. W. Hanneman, C. F. Spencer, and J. F. Johnson, Anal. Chem., 32:1386 (1960).

69.   W. W. Hanneman, J. Gas Chromatog., 1:18 (1963).

70.   P. W. Solomon, Anal. Chem., 36:476 (1964).

71.   G. R. Schultze and W. J. Schmidt-Küster, Z. Anal. Chem., 170:232 (1959).

72.   R. M. Barrer and L. V. C. Rees, Trans. Faraday Soc., 57:999 (1961).

73.   L. D. Belyakov, A. V. Kiselev, N. V. Kovaleva, L. N. Rozanova, and V. V. Khopina, Zh. Fiz. Khim., 42:77 (1968).

74.   F. T. Eggertsen, H. S. Knight, and S. Graenings, Anal. Chem., 28:303 (1956).

75.   W. Schneider, H. Bruderreck, and I. Halasz, Anal. Chem., 36:1533 (1964).

76. J. Janak, Collection Czech. Chem. Commun., 18:798 (1953); 19:684 (1953).
77. A. V.Kiselev and Ya. I. Yashin, Neftekhimiya, 4:494 (1965).
78. S. P. Zhdanov, V. I. Kalmanovskii, A. V. Kiselev, M. M. Fiks, and Ya. I. Yashin, Zh. Fiz. Khim., 36:1118 (1962).
79. M. Mohnke and W. Saffert, in collection: Gas Chromatography 1962, M. van Swaay, (ed.), London, 1962, p. 214.
80. F. Bruner and G. P. Cartoni, J. Chromatog., 18:390 (1965).
81. I. Halasz and C. Horvath, Nature, 197:71 (1963).
82. R. D. Schwartz, D. J. Brasseaux, and G. R. Shoemaker, Anal. Chem., 3 :496 (1963).
83. C. Horvath, Trennsäulen mit dünnen porösen Schichten für die Gaschromatographie, Dissertation, Frankfurt am Main, 1963.
84. I. Halasz and E. Heine, Nature, 194:971 (1962).
85. H. V. Carter, Nature, 197:684 (1963).
86. I. Halasz and F. Holdinghausen, in collection: Advances in Gas Chromatography, A. Zlatkins, New York, 1967, p. 23.
87. O. L. Hollis, Anal. Chem., 38:309 (1966).
88. A. V. Kiselev, I. A. Migunova, and Ya. I. Yashin, in collection: Gas Chromatography, Moscow, Izd. NIITEKhim, Issue 6, 1967, p. 84.
89. C. Vidal–Madjar and G. Guiochon, Bull. Soc. Chim. France, 1966, p. 1096.
90. C. Vidal–Madjar and G. Guiochon, Separation Science, 2:155 (1967).
91. C. Vidal–Madjar and G. Guiochon, Compt. Rend. Acad. Sci., Paris, 265:26 (1967).
92. C. Vidal–Madjar and G. Guiochon, Nature, 215:1372 (1967).
93. C. Vidal–Madjar and G. Guiochon, Proceedings of the Sixth Symposium on Gas Chromatography, Berlin, 1968 (in press).
94. S. T. Sie and G. W. A. Rijnders, Anal. Chim. Acta, 38:31 (1967).
95. S. T. Sie, J. P. A. Bleumer, and G. W. A. Rijnders, in collection: Preprints of SeventhInternational Symposium on Gas Chromatography, Copenhagen, June 25-28, 1968, C. L. A. Harbourn (ed.), London, Institute of Petroleum, 1968, Paper 14.
96. G. C. Goretti, A. Liberti, and G. Nota, J. Chromatog., 34:96 (1968).

Chapter II

Role of the Surface Chemistry of Adsorbents and Nature of Molecular Interactions

NONSPECIFIC AND SPECIFIC

MOLECULAR INTERACTIONS

The effects of geometrical inhomogeneity in the surfaces and pores may be largely avoided by using crystalline adsorbents, and also amorphous ones with large pores with surfaces modified either chemically or by adsorption. The surface chemistry of the adsorbent will then largely control the adsorption and selectivity of the gas-adsorption column. The surface chemistry determines the nature and energy of the interactions with the adsorbate. Interaction with a homogeneous surface and the state of adsorbed molecules on such a surface (gas-adsorption chromatography) allow more readily of theoretical treatment than do the interactions on a solution in a liquid film (gas–liquid chromatography). All the molecules in a solution are mobile and so interact with fluctuating centers, whereas the molecules adsorbed on a reasonably smooth solid surface interact mainly with the nearest force centers, which are fixed.

Molecular interactions have a single quantum-mechanical basis, but the theory is poorly developed, and no general expression is available for the interaction potential at short distances; this potential is therefore usually represented as the sums of apparently independent interactions: dispersion, electrostatic, repulsive, and chemical. The surface and adsorbate may give rise to various interactions, which range from essentially molecular (nonspecific and specific), with the chemical individuality of the partners retained, to chemical, with the individuality lost and a new surface compound formed.

Nonspecific interaction is universal and occurs between any partners; it is largely of dispersion-force origin. Specific interaction is related to details of the electron-density distribution at the edges of the molecules, such as details of the individual bonds or links. These interactions in general are not classical; only in the limit of large distances do they reduce to classical electrostatic interactions. The hydrogen bond is a specific interaction of molecular type. Closer interactions, e.g., with complete charge transfer (as in donor−acceptor coordination bonds), lead to loss of chemical individuality of the interacting partners. Characteristic examples of such an interaction occur in the adsorption of anthracene on an aluminosilicate catalyst [6], a cation-free zeolite [7], and aluminum dioxide [8], where ESR occurs.

This classification of interactions is partly arbitrary, because intermediate cases occur; but it assists in giving order to a variety of facts in a general qualitative way.

High selectivity for weak interactions is important in chromatography, so here molecular interactions (nonspecific and specific) are of interest, not chemical ones.

The terms specific and nonspecific (as applied to molecular interactions) are convenient in classifying similar interactions, though one would expect that further development of the theory either should give each a more detailed physical content or should lead to replacement by more rational terms [9, 10].

CLASSIFICATION OF MOLECULES
AND ADSORBENTS BY CHARACTER
OF MOLECULAR INTERACTION

It is convenient to classify together some typical similar types of interaction by relating them to molecular features, as has been done [1-5] for molecular (nonchemical) interactions by Kiselev in 1963-5. It is convenient to divide molecules into groups A-D (Table 1) by reference to the peripheral electron-density distribution.

Group A. The simplest case as regards electronic structure is that of spherically symmetrical shells, as in the inert gases. These molecules retain their individuality (do not interact chemically) and can interact with other molecules only nonspecifically,

Table 1. Classification of Molecules and Adsorbents by Capacity for Nonspecific and Specific Interaction [3-5]

Molecules	Adsorbents		
	Type I: without ions or active groups (graphitized carbon blacks, BN, surfaces bearing only saturated groups)	Type II: with localized positive charges (acid OH, exchangeable cations of small radius)	Type III: with localized negative charges (esters, nitriles, carbonyl groups, exchangeable anions of small radius)
Group A: with spherically symmetrical shells or σ-bonds (inert gases, saturated hydrocarbons)	Nonspecific interactions governed mainly by dispersion forces		
Group B: electron density locally concentrated on bonds or links, π-bonds (N$_2$, unsaturated and aromatic hydrocarbons) and lone electron pairs (esters, ketones, tertiary amines, nitriles, etc.)	Nonspecific interactions	Nonspecific + specific interactions	
Group C: with positive charge localized on peripheral links (e.g., some organometallic compounds)			
Group D: with functional groups having locally concentrated electron density and positive charge on adjacent links (molecules with OH and NH groups)			

mainly via the universal dispersion forces, not via details of the local electron-density distribution associated with particular bonds or atoms. To this group we may assign the saturated hydrocarbons, which have only σ-bonds between the carbon atoms; these have no elongated orbitals, i.e., there are no peripheral concentrations of electron density.

Group B. Here the electron density (negative charge) is localized at the periphery of particular links. The localization is at the periphery of the bonds in unsaturated and aromatic hydrocarbons, and in all other such molecules having π-bonds, e.g., nitrogen. This density localization occurs on links in functional groups having nonbonding pairs, e.g., oxygen in water, alcohols, esters, and ketones, or nitrogen in ammonia, amines, and cyanides, or sulfur in organic sulfides.

This localization provides the basis for specific interactions, which are still molecular (not chemical), though this can occur only if the other partner has localized positive charge, e.g., peripheral H atoms of acid type or peripheral cations, especially exchangeable ones of small radius.

The O and N in water, alcohols, and amines (primary and secondary) have nonbonding pairs in the OH, NH_2, or NH groups, together with partial protonization of the H, which can also participate in specific interaction. This complicates the molecular interaction; in particular, it provides the basis for association, and so it is best to place in group B only molecules having solely localized electron density, i.e., π-bonds (nitrogen, unsaturated and aromatic hydrocarbons) and nonbonding pairs (esters, ethers, ketones, tertiary amines, cyanides, sulfides). These interact nonspecifically with molecules of group A. Group B molecules differ from those of group A in that they can interact specifically with positive charges localized on molecules or surfaces; e.g., they can form H bonds to OH and NH_2 groups. However, they show no strong specific interaction one with another, although the dipoles or quadrupoles may make some contribution to the total interaction energy.

Group C. Here positive charge is localized on small links if the corresponding excess electron density is distributed over adjacent links (i.e., if there is no adjacent link with peripherally localized electron density, as in OH, NH_2, and NH). The molecules of many organometallic compounds are probably of this type. The interaction with molecules of group A is nonspecific, while that with group B molecules is specific.

Group D. In this group we have molecules with adjacent small links, one bearing positive charge and the other with peripheral electron density, e.g., as in OH, NH_2, and NH groups. Examples are water, alcohols, and primary and secondary amines (not tertiary). The interaction with group A is nonspecific; that with groups B and C is specific, as is that within the group.

Many functional groups are of complex structure, as in COOH or ester groups, which contain hydroxyl or alkoxyl oxygen together with carbonyl oxygen, each of these being capable of specific interaction.

Many molecules, especially large ones, contain several separated links (perhaps of different types) that are capable of specific molecular interaction, as in amino-alcohols, amino-acids, and amino-esters.

Nonspecific interaction occurs in all cases, while specific interaction occurs additionally only when the interacting partners belong to groups B, C, and D.

Particular interest attaches to molecules with highly polarizable bonds and links, and also to molecules with conjugated bonds, such as divinyl, many benzene derivatives, polynuclear aromatics, and heterocyclics such as pyridine, furan, thiophen, and tetracyanoquinodimethane:

$$N{\equiv}C{\diagdown}C{\equiv}N$$
$$C{=}C{\Big\langle}{\Big\rangle}C{=}C$$
$$N{\equiv}C{\diagup}C{\equiv}N$$

Consider now the interactions of these groups with the various adsorbents. The surface of an adsorbent may be considered in the way discussed above for molecules. It is convenient to classify [2-5] adsorbents into basic types in accordance with the charge distribution at the surface.

Type I, Nonspecific. The surface bears no functional groups or exchangeable ions; examples are graphitized thermal carbon black, BN, and saturated hydrocarbons, especially polymers such as polyethylene. These interact largely nonspecifically with molecules of all four groups.

Type II, Specific, with Localized Positive Charge. The surface interacts specifically with any molecule having peripherally localized electron density (groups B and D); these are especially adsorbents bearing acidic OH groups, such as hydroxylated surfaces of acid oxides, in particular silica. The d-shell vacancy in the silicon atom here leads to a shift in electron density such as partly to protonize the OH hydrogen [11]. However, a similar charge distribution can occur in the absence of acidic OH groups, provided that the positive charge is displaced to the surface and is localized on a particle of small radius, while the negative charge is distributed over a much larger volume. Hence type II includes adsorbents whose surfaces bear aprotonic acid centers or cations of small radius, with the negative charge distributed over the internal bonds of some large complex anion. Zeolites are examples of such adsorbants; here the positive charge is localized in exchangeable cations, while the negative charge is distributed over the $(AlO_4)^-$ anions of the zeolite framework [12-14].

Type II thus covers specific adsorbents bearing surface-localized positive charges as partly protonized H atoms of surface OH groups or as cations balanced by large (especially complex) anions.

The specificity of the molecular interaction of an ionic adsorbent is [15] reduced when the cations and anions on the faces become comparable in size and when the ion-exchange capacity is lost.

Type III, Specific, with Localized Negative Charge. Surfaces of this type are readily produced by deposition on a nonspecific adsorbent (i.e., type I, such as graphitized carbon) of close packed monolayers of a group B substance (e.g., polyethylene glycol [5, 16]) or by formation of function groups (e.g., CN) by chemical modification [5].

Hence the charge distribution allows us to distinguish one nonspecific type and two specific types.

Some adsorbents produce various local interactions on account of the presence of a variety of functional groups and conjugated bonds, e.g., a layer of Cu phthalocyanin on graphitized thermal carbon black [142], 143]. The parts of the surface have the following structure:

Table 1 [3-5] classifies molecules and adsorbents in this way in terms of interactions. Here we always envisage molecular interaction, not chemical, and hence molecular (physical) adsorption, in which the interacting partners do not lose their chemical individuality.

TYPE I ADSORBENTS

Graphitized Thermal Carbon Black. This is an adsorbent with a highly uniform surface, e.g., as when graphitized at ~3000°C, when the behavior is almost completely unspecific and the specific surface is small (6-30 m^2/g). Figure 1 shows that the particles are polyhedra with planar faces, which are external basal faces of graphite crystals [17, 18], so the surface is exceptionally homogeneous.

Adsorbate –adsorbate interactions are detectable, on account of the high surface homogeneity of this carbon; the differential heat of adsorption increases as the surface becomes coated, and, if this increase is fairly large [19], it leads to curved adsorption isotherms, which initially are convex to the pressure axis but then have an inflection, which corresponds to a degree of filling $\theta \approx 0.5$.

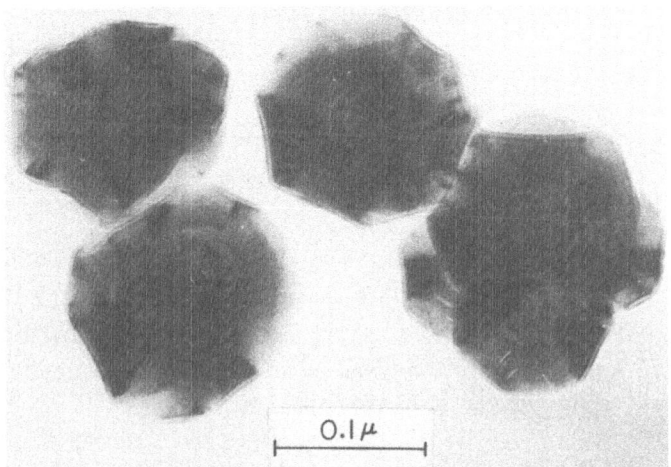

Fig. 1. Electron micrograph of Sterling MT, 3100°, D4 graphitized thermal carbon black by Cabot(recorded by E. B. Oganesyan).

A satisfactory description is given for the filling of a homogeneous surface as a function of p [19-23, 144-147] and T [5, 24,
25, 145, 147] by equations that take into account the adsorbate− adsorbent and adsorbate−adsorbate interactions [19-23, 144-147]. The
simplest of these is that given by Kiselev [19]:

$$p = \frac{\theta}{K_1(1-\theta)(1+K_n\theta)} = \frac{\theta}{A_1(1-\theta)\exp(B_1/T)[1+A_n\theta\exp(B_n/T)]} \cdot \text{(II, 1)}$$

Here K_1 is the Henry constant (equilibrium in adsorbate−adsorbent
interaction), K_n is the equilibrium constant of the adsorbate−ad-
sorbate interaction [19], and A_1, B_1, A_n, B_n are constants whose
numerical values are deduced from experimental curves [24, 25].
Figure 2 shows the adsorption of ethane on graphitized carbon black,
which is described by this equation.

Hill's equation [21] is also often used, which describes two-
dimensional condensation at temperatures below the critical value
for a two-dimensional adsorbed layer. The Wilkins equation [144]
is more general; it corresponds to the equation of state for a two-
dimensional layer with virial coefficients and expresses the p dependence of the adsorption uptake a (per g of adsorbent) or the surface concentration $\alpha = a/s$ (per m^2, in which s is the specific sur-

face of the adsorbent) as follows [144-147]:

$$p = \alpha \, \exp \, (C_1 + C_2 \, \alpha + C_3 \, \alpha_2 + \ldots) \qquad \text{(II, 1a)}$$

in which C_1, C_2, C_3, etc., are constants dependent on temperature, with exp (C_1) representing Henry's constant. Correspondingly, the heat of adsorption is

$$Q = Q_1 + Q_2 \, \alpha + Q_3 \, \alpha^2 + \ldots \qquad \text{(II, 1b)}$$

If we neglect the dependence of Q on T, we have $C_i = B_i - Q_i/RT$, in which the B_i are constants independent of T. The dependence of α on p and T is then given [145, 147] by

$$p = \alpha \exp \, (B_1 + B_2 \, \alpha + B_3 \, \alpha^2 + \ldots) \exp \left(- \frac{Q_1 + Q_2 \, \alpha + Q_3 \, \alpha^2 + \ldots}{RT} \right) \quad \text{(II, 1c)}$$

These series converge rapidly, the first four terms being adequate to describe the dependence of α on p and T, and of Q on α, up to $\theta \approx 0.7$ [147]. For α fairly small, (II, 1a) for simple compounds becomes the virial series

$$p = K_1 \, \alpha + K_2 \, \alpha^2 + \ldots \qquad \text{(II, 1d)}$$

used in the static theory of adsorption for low θ and high T. Equations (II, 1a), (II, 1c), and (II, 1d) have an advantage over (II, 1) in that they do not contain α_m or a_m, the capacity of a monolayer, which is needed to determine $\theta = \alpha/\alpha_m = a/a_m$ from the data on a, since α_m or a_m cannot be determined accurately, especially for adsorbents with fine pores.

Equations (II, 1) and (II, 1c) imply that the chromatographic bands (and also the adsorption isotherms) are complex for low T and high θ; these equations tend to Henry's equation for the adsorption isotherm on a homogeneous surface in the absence of adsorbate−adsorbate interaction as T increases or θ decreases. The relatively small slope of the Henry isotherms for large T, i.e., the small $K_1 = A_1 \exp (B_1/T)$, corresponds to rapid desorption, which provides narrow and symmetrical peaks. Barrer and Rees [26] have shown that the actual isotherms approximate to Henry ones even for inhomogeneous surfaces at sufficiently high T.

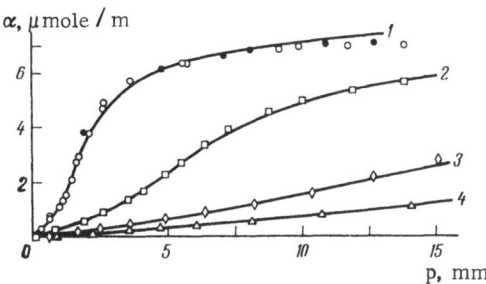

Fig. 2. Adsorption isotherms for ethane on Sterling
FT, 2800° carbon black (Cabot) at: 1) −100°; 2)
−85°; 3) −70°; 4) −60°C. The points were recorded
by the static method; the curves are from (II, 1).

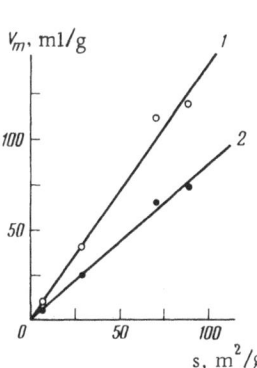

Fig. 3. Relation of V_m for
graphitized carbon black
to s for: 1) pentane; 2)
diethyl ether (75°C).

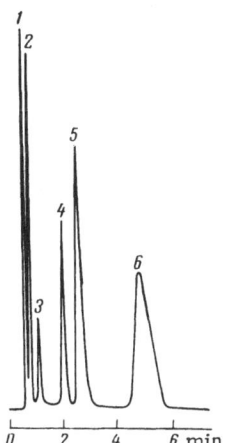

Fig. 4. Chromatogram on
graphitized carbon black
from a mixture containing
compounds of groups A, B,
and D (column 100 × 0.5
cm, 235°C): 1) cyclohex-
ane; 2) cyclohexanol; 3)
aniline; 4) nitrobenzene;
5) acetophenone; 6) n-
decane.

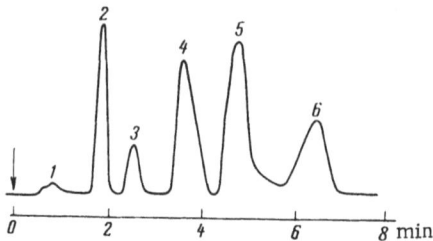

Fig. 5. Chromatogram on a 100 × 0.5 cm column
of graphitized thermal carbon black at 93°C, carrier gas He, 60 ml/min, katharometer: 1) water;
2) cyclohexane; 3) cyclohexene; 4) benzene; 5) ˙
pyridine; 6) n-hexane.

The properties of unit area of graphitized carbon black are
highly reproducible and are virtually independent of the specific
surface s. Figure 3 shows that the specific retained volume V_m
(per gram carbon) is proportional to s at small θ for carbon graphitized at ~ 300°, i.e.,

$$V_m = V_s \cdot s, \tag{II, 2}$$

in which $V_s = V_m/s$ is the volume adsorbed [27, 28] per unit surface (outer basal face of semiinfinite graphite lattice). This quantity represents K_1, Henry's constant. The V_s are reproducible
physicochemical constants that characterize given adsorbate–adsorbent systems at given T.

The V_s may be used for identification in chromatography and
for rapid determination of s.

Gas-Adsorption Chromatography on Graphit-
ized Carbon Black. Figure 4 shows a chromatogram of a mixture of substances belonging to groups A, B, and D, i.e., ones in-
capable and capable of specific interaction: cyclohexane, cyclohexanol, aniline, nitrobenzene, acetophenone, and n-decane [4, 29].
Some of these have the very active groups OH, NH_2, and CO, which
give strong specific interactions, or NO_2 (very large dipole moment),
but n-decane is the last to appear, because the graphitized carbon
black has no sites with localized charge that could give a specific
interaction.

Figure 5 shows a chromatogram from a mixture of six-membered ring compounds, planar and nonplanar, with the straight-

chain n-hexane [4, 30]; here again the local details of the electronic structure do not make themselves felt. For example, the heats of adsorption are similar for pyridine (nonbonding electron pair, large peripheral dipole moment, π-bonds in ring) and benzene (only π-bonds) [30, 31]. The heats of adsorption of cyclohexane, cyclohexene, benzene, and pyridine are less than that of n-hexane, although the molecular weights are similar but the boiling point of n-hexane is much lower. This sequence of heats corresponds to the sequence of nonspecific (mainly dispersion) interactions (see [2, 5] for literature). The order of release of these six-membered compounds is due mainly to differences in spatial structure. There is only nonspecific interaction with the adsorbent, which is almost independent of the dipole or quadrupole moment and of the local charge distribution in the bonds or links.

This is clear from Table 2 [3, 4, 32], which gives the number of carbon atoms n, the molecular weight M, the boiling point T, the dipole moment μ, total polarizability α, V_s for graphitized carbon black (proportional to the absolute values of the Henry constants) at several temperatures, and Q_0 (initial differential heat of adsorption).

Table 2 shows that V_s increases with α, which defines the energy of nonspecific dispersion interaction.

Table 2 also shows that there is no direct relation between this V_s and T, μ, or M. The reason is that T is related to molecular interaction in the liquid, where association may occur for group D molecules, whereas no such association can occur on carbon black in the initial stages of adsorption; and μ makes no considerable contribution to Q_0 [33], although it causes some polarization of the atoms in the graphite, and so hardly affects V_s. The effect of M is largely masked by differences in molecular geometry, which have a very pronounced effect on the orientation and adsorption energy. The nonspecific dispersion interaction is determined by the polarizability and magnetic susceptibility of the individual links, as well as by the orientation of the links relative to the surface.

The V_s, or the corresponding relative values as ratios to the n-alkanes, $V_m/V_{m. n-alkane}$, may be used for identification; similarly, V_s can give valuable information on the geometrical structure of an unknown compound. Figures 6 and 7 show that log V_s

Table 2. Approximate Values of Retained Volume V_s and Initial Differential Heat of Adsorption Q_0 for Graphitized Carbon Black (n = number of carbon atoms in molecule, M = molecular weight, T = boiling point, μ = dipole moment, α = total polarizability)

n	Compound	M	T, °C	μ, D	α, Å³	v_s, ml/m²				Q_0, kcal/mole
						50° C	100° C	150° C	200° C	
1	Methanol	32.0	64.7	1.67	3.23	—	0.06	0.03	—	5.3
	Chloroform	119.4	61.2	8.23	8.23	3.8	0.56	0.13	—	8.9
	CCl₄	153.8	76.8	0.0	10.5	6.3	0.83	0.19	0.058	9.6
2	Ethanol	46.1	78.4	1.7	5.06	0.28	0.14	0.05	0.03	6.9
	Acetic acid	60.0	118	1.4	5.03	—	0.46	0.16	0.06	8.3
	Dichloroethane	99.0	83.7	1.42	—	5.25	0.48	0.105	0.048	7.3
3	n-Propanol	60.1	97.8	1.66	6.89	—	0.28	0.09	0.04	8.1
	Isopropanol	60.1	82.4	1.65	—	0.95	0.25	0.09	—	7.0
	Acetone	58.1	56	2.73	6.32	—	0.28	0.05	—	8.3
	Propyl chloride	78	40	2.15	—	2.76	0.43	0.10	—	9.5
4	Dioxane	88.1	101.4	0.45	—	—	0.65	0.15	0.05	9.8
	t-Butanol	74.1	82.9	—	—	2.1	0.42	0.12	—	8.3
	s-Butanol	74.1	99.5	—	—	4.3	0.76	0.20	—	8.9
	Isobutanol	74.1	107	1.6	—	4.3	0.76	0.20	—	8.9
	n-Butanol	74.1	117	1.66	8.72	6.95	1.1	0.26	—	9.4
	Diethyl ether	74.1	34.5	1.17	9.02	—	0.44	0.13	—	8.6
	Thiophen.	84	84	0.63	—	—	0.84	0.20	0.065	9.0
5	n-Valerianic acid	102	186	—	—	—	10	1.38	0.29	12.3
	n-Amyl alcohol	88	138	1.8	10.55	—	5.0	0.87	0.24	10.7

Table 2 (Continued)

n	Compound	M	T, °C	μ, D	α, Å³	v_s, ml / m²				Q_0, kcal/mole
						50° C	100° C	150° C	200° C	
5	Pyridine	79.1	115.4	2.2	9.5	21.3	2.1	0.47	—	10.1
	Furfural	96.1	162	—	—	—	4.8	1.00	0.27	10.6
	n-Pentane	72.1	36.3	—	9.95	—	0.76	0.20	0.08	8.9
6	Cyclohexane	84.2	81	0.61	10.9	6.18	0.83	0.24	—	8.7
	Cyclohexene	82.2	83	0	—	11.2	1.2	0.3	—	9.1
	Benzene	78.1	80.1	0	10.32	17.2	1.7	0.41	0.12	9.8
	n-Hexane	86.2	69	—	11.78	27.5	3.2	0.63	—	10.4
	Cyclohexanol	100	160	—	—	—	3.67	0.68	0.18	11.2
	Cyclohexanone	98.1	155.6	2.8	—	—	3.67	0.68	0.18	11.2
	Phenol	94.1	181.4	1.4	—	—	13.3	1.82	0.40	13
	Aniline	93.1	184.4	1.48	—	—	20.2	2.74	0.61	13
	Nitrobenzene	123	210.9	4.21	12.92	—	62	9.5	1.5	14.2
	n-Hexanol	102	157	—	—	—	15	3.4	0.57	12
	α-Picoline	93	129.4	1.72	—	—	7.35	3.00	0.37	11.3
	β-Picoline	93	144.1	2.30	—	—	8.85	3.1	—	11.5
	Chlorobenzene	112.6	131.2	1.72	12.25	—	9.2	1.87	0.52	10.8
	Bromobenzene	157	156	1.75	—	—	14.6	2.87	0.86	11.0
	Iodobenzene	204.2	188.6	1.60	—	—	35	6.3	1.68	11.4
	Hydroxyquinone	110	286	2.47	—	1350	83.2	11.2	2.19	12.6

Table 2 (Continued)

n	Compound	M	T, °C	μ, D	α, Å³	v_s, ml/m²				Q_0, kcal/mole
						50° C	100° C	150° C	200° C	
7	Benzoic acid	122.1	250	—	—	—	12.3	1.77	0.42	11.9
	Toluene	92.1	110.8	0.36	12.15	—	10.1	1.65	0.39	11.6
	n-Heptane	100.2	79.1	0	13.61	—	12.0	1.9	0.45	12.5
	n-Heptanol	116	175	1.70	—	—	69.5	7.95	1.41	13.2
	m-Cresol	108	202	1.54	—	—	68	9.5	2.1	12.5
8	Ethylbenzene	106	136.8	—	—	40.7	47.8	4.6	0.83	12.7
	m-Xylene	106	139.1	0.34	14.2	—	64.5	17	3.2	16.2
	n-Caprylic acid	144.2	237.5	—	—	—	710	44.5	5.4	16.8
	n-Capryl alcohol	130	194	1.67	—	—	70	22.4	3.5	14.6
	Dimethylaniline	120	164.6	1.61	—	3980	288	39.8	8.3	14.0
	Acetophenone	120	202	3.01	—	—	158	38.0	6.9	13.0
	n-Octane	114	125.7	0	—	—	42.6	6.02	1.34	13.4
9	Isopropylbenzene	120	152.4	0.40	—	725	47.9	6.6	1.32	13.2
	Propylbenzene	120	159	0.40	—	964	86.3	9.34	2.34	14.3
	Mesitylene	120	164.7	0.07	—	131.8	23.9	8.38	2.5	15
	n-Nonane	128	150	—	—	—	190.5	20.0	3.45	14.8
	n-Nonanoic acid	158	254	—	—	—	1780	151	11.5	18.3
	n-Nonanol	144	213.5	1.60	—	—	1260	79.4	10	16.0
10	t-Butylbenzene	134	168	0.53	—	—	832	11.0	2.2	13.8
	s-Butylbenzene	134	171	0.42	—	—	139	11.5	1.95	15.0

Table 2 (Continued)

n	Compound	M	T, °C	μ, D	α, Å³	v_s, ml/m²				Q_o, kcal/mole
						50° C	100° C	150° C	200° C	
10	Isobutylbenzene	134	182	0.37	—	—	185	16.6	2.85	15.2
	n-Butylbenzene	134	183.3	0.42	—	—	294	24.7	3.95	15.3
	Durene	134	191.2	—	—	—	832	158.5	19.0	18.9
	n-Decane	144	174	0	—	—	797	62.0	8.20	16.7
	Naphthalene	128.2	218	—	—	—	1950	142	17.9	17.3
	n-Decyl alcohol	158.3	232.9	1.63	—	—	617	79.4	15.5	17.4
	Fe·ocene	186	249	—	—	—	38	5.1	1.04	12.8
	Camphene	136	160	—	—	89.4	13.9	2.32	0.59	11.8
	α-Pinene	136	156	0.8	—	100	14.8	2.43	0.60	12.1
	Camphor	152	203	3.0	—	158	23.2	3.71	0.9	12.2
	Borneol	154	212	1.56	—	230	31	4.63	1.06	12.6
	Fenchone	152	194	2.9	—	295	37.3	6.7	1.16	13.0
	Fenchol	154	201	—	—	309	39	6.8	1.19	13.0
	Δ³-Carene	136	170	—	—	782	81.7	9.55	1.82	14.2
	Sabinene	136	165	—	—	2140	197	21.2	3.68	14.8
	Limonene	136	176	0.5	—	2140	197	21.2	3.68	14.8
	p-Cymene	134	177	—	—	2140	197	21.2	3.68	14.8
	α-Terpineol	154	220	—	—	331.0	270	26	4.1	15.5
	Menthone	154	207	2.8	—	3550	316	32.4	5.43	15.0
	Menthol	156	215	1.57	—	4880	400	40.3	6.53	15.0
	Carvone	150	230	2.8	—	8430	606	52	7.5	16.7
11	n-Undécane	156	196	0	—	56300	4020	194	22.2	18.0

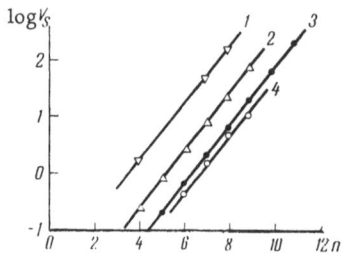

Fig. 6. Relation of log V_s at 150°C to n for graphitized thermal carbon black: 1) normal acids; 2) normal alcohols; 3) normal alkanes; 4) normal alkylbenzenes.

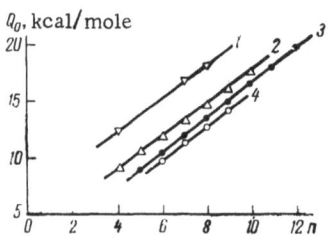

Fig. 7. Relation of Q_0 to n for the compounds of Fig. 6.

and Q_0 have linear relations to n for homologous series, so unknown components can be identified and V_s within a homologous series predicted.

Figure 7 shows that the Q_0 of some derivatives of n-alkanes are higher than those for n-alkanes with the same number of carbon atoms n. This difference is 1.5 kcal/mole for the n-alcohols, and 2.9 kcal/mole for the normal fatty acids (as measured at 50-200°C, so they are somewhat less than the Q_0 at 20°C). Again, the n-alkylbenzenes have Q_0 less than those of the corresponding n-alkanes with the same n by 0.7 kcal/mole [5, 32].

The Q_0 for n-alkanes and unbranched derivatives are additive by links, the increment being as follows: $Q_{CH_3} = 2.1$; $Q_{CH_2} = 1.6$; $Q_{OH} = 2.1$; $Q_{O\,ethers} = 1.3$; $Q_{COOH} = 5.2$ kcal/mole (also for 50-200°C). These values give an approximate Q_0 for compounds containing these groups.

Figure 8 shows isothermal (150°C) results for $\Delta\mu_0$ (change in chemical potential of adsorbate, proportional to log V_s) and ΔS_0 (standard change in differential entropy) as functions of Q_0. The points for the n-alkanes (group A) lie on straight lines [32]. These relationships are discussed in more detail in chapter IV; here we only note that the points for other organic compounds containing various bonds and functional groups (molecules of groups B and D) lie near these lines.

These examples show that graphitized carbon black is a convenient material for separating compounds by reference to the energy of the nonspecific dispersion interaction, which is almost independent of the group to which the molecule belongs.

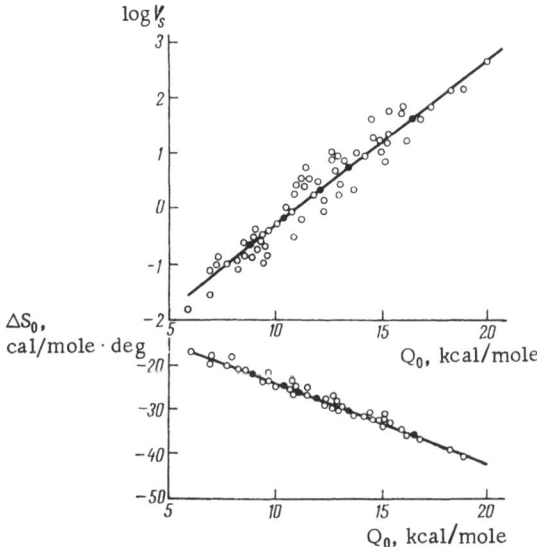

Fig. 8. Relation of log V_s and ΔS_0 to Q_0 for: ●) n-
alkanes; O) derivatives of these (Table 2).

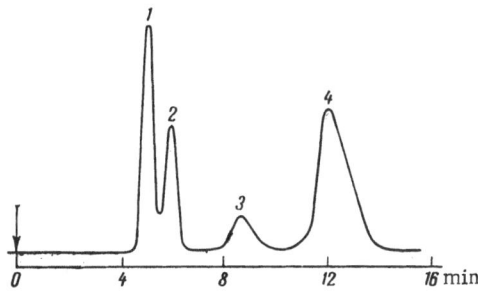

Fig. 9. Chromatogram of the butylbenzenes at 208°C
on a column 100 × 0.5 cm of graphitized thermal
carbon black (helium flow rate 40 ml/min, katharometer): 1) t-butylbenzene; 2) s-butylbenzene; 3) i-butyl-
benzene; 4) n-butylbenzene.

In particular, the adsorbent is very convenient for group D
compounds, such as water, monohydric and dihydric alcohols, organic acids, and amines, which are difficult to separate on other
adsorbents. Chapter V gives some examples of the use of graphitized carbon blacks in analysis.

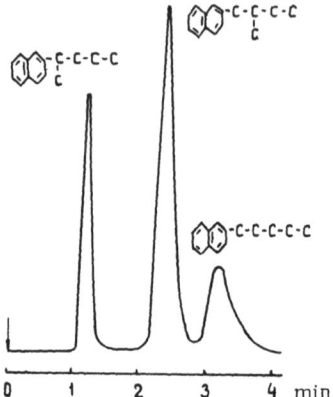

Fig. 10. Chromatogram of the amylnaphthalenes at 300°C on a column 50 ×0.3 cm of graphitized thermal carbon black, s = 8.1 m²/g (nitrogen flow rate 30 ml/min).

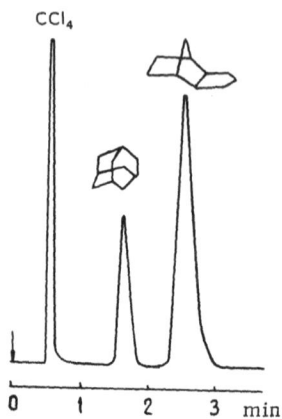

Fig. 11. Chromatogram of a mixture of: 1) adamantane and 2) tetrahydrodicyclopenta-diene at 124°C on a column as in Fig. 10; nitrogen flow rate 40 ml/min.

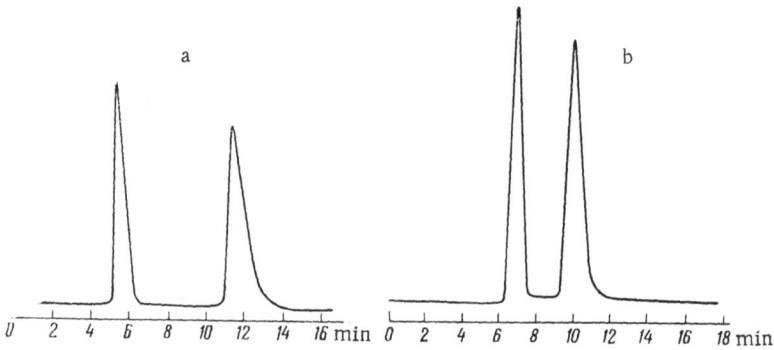

Fig. 12. Chromatogram on a column 100 × 0.5 cm of graphitized thermal carbon black (s = 7.6 m²/g, N₂ 50 ml/min, flame ionization detector): a) cis and trans isomers of 1,4-di-t-butylcyclohexane at 250°C; b) cis and trans isomers of normal 1-methyl-3-hexylcyclohexane at 300°C.

Separation of Isomers on Graphitized Carbon Black. The energy of nonspecific interaction is very much dependent on the distance between the surface and the force centers of the links in the molecule, so the planar surfaces of graphitized carbon may conveniently be used to separate substances that differ only in geometrical structure. Figure 9 illustrates this for the

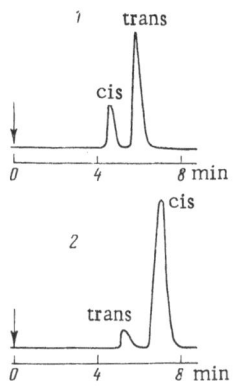

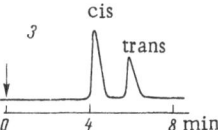

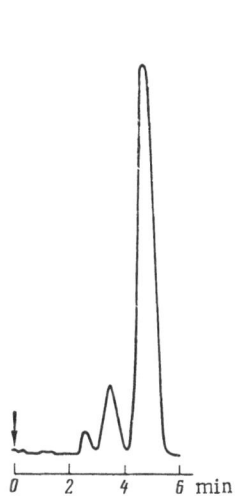

Fig. 13. Chromatogram
on a column 200 × 0.5
cm of graphitized ther-
mal carbon black for
cis and trans decalins
at 210°C (carrier gas 50
ml/min, katharometer).

Fig. 14. Chromatogram on
a column 200 × 0.5 cm of
graphitized thermal carbon
black at 170°C for mixtures
of cis and trans isomers of
methylcyclohexanols: 1)
1,2-; 2) 1,3-; 3) 1,4- (helium 32 ml/min, katharometer).

isomeric butylbenzenes; n-butylbenzene has the largest number of
side-chain links that can be in direct contact with the basal face of
the graphite, while t-butylbenzene has the least; hence these appear respectively last and first, while s-butylbenzene and i-butyl-
benzene run between these, in accordance with the geometry and
possible orientations of the molecules [30, 34]. The isomeric amylnaphthalenes are eluted in the same order (Fig. 10).

The sequence of isomeric butenes also shows marked effects
on V_s and Q_0 from the molecular geometry. In particular, the selectivity in separation of but-1-ene and isobutene on graphitized
thermal carbon black is higher than for other known adsorbents
[35].

Separation on graphitized carbon black becomes difficult if
different compounds present equal numbers of carbon atoms to the
surface; e.g., the larger van der Waals radius of CH_3 (2.0 Å) relative

to that of ring CH (1.8 Å) causes only one CH group in m-xylene to touch the plane, so m-xylene is more weakly adsorbed than the o- and p-isomers; it emerges first. The o- and p-isomers have two CH groups touching the surface, so they have similar retention times; but they are easily separated on specific adsorbents after removal of the m-xylene, because their electron-density distributions are very different.

One of the most interesting examples of two isomers differing in geometrical structure is that of adamantane and tetrahydrodicyclopentadiene, which are not separated in a gas–liquid column with Apiezon L, whereas a relatively small column of graphitized carbon black provides a very good separation (Fig. 11).

Separation of cis and trans isomers is very important; Fig. 12 illustrates this for alkylcyclohexanes, and Fig. 13 for the decalins. The latter has three stable isomers [36]: boat–boat cis, chair–chair cis, and chair–chair trans, which should leave the column in that order on the basis of their geometry and possible orientation. Three peaks are observed [30, 34]. Figure 14 illustrates the performance for the 1,2-, 1,3-, and 1,4-methylcyclohexanols [30, 34].

The naphthene fraction of gasoline contains 1,4-dimethylcyclohexane and 1,3-dimethylcyclohexane, which are difficult to separate with liquid phases. The V_S for graphitized carbon black for the stereoisomers (cis and trans) of these range from 7 to 20 ml/m^2 at 100°C (Table 3) [148]; each hydrocarbon emerges as two peaks, as seen in Fig. 15, the first peak for 1,4-dimethylcyclohexane corresponding to the cis isomer, while the second represents the trans one. The order is reversed for 1,3-dimethylcyclohexane. The separation on this material at 100°C (Fig. 15) is complete, whereas trans-1,3-dimethylcyclohexane cannot be separated from cis-1,4-dimethylcyclohexane (which has the same boiling point) with liquid phases.

The cis isomers of some unsaturated compounds have lower t_R, with higher boiling points, than the trans isomers. It has been shown [149] that olefins of cis configuration have lower t_R than those of trans configuration. For instance, the relative retained volume V_m (relative to n-pentane) for a capillary column of graphitized carbon black with cis-but-2-ene (b.p. 3.7°C) is 0.156, while for trans-

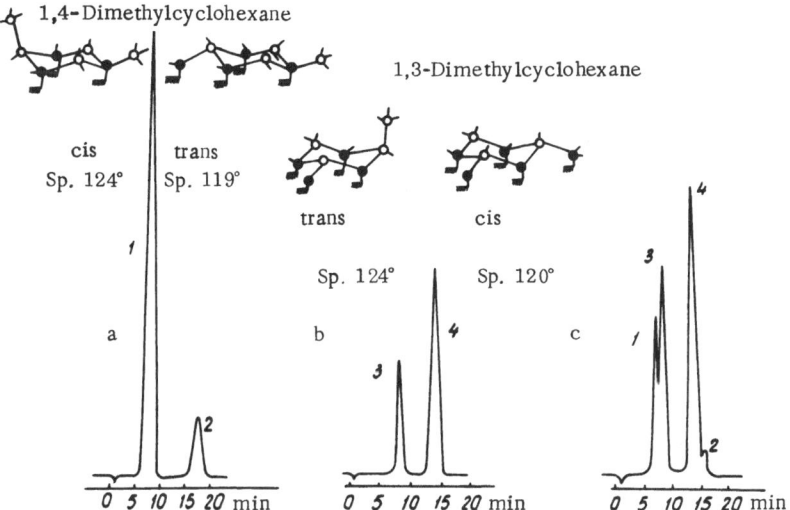

Fig. 15. Stereoisomers on a 120 × 0.25 cm column of graphitized carbon black (s = 7.6 m²/g) at 100 °C: a) cis and trans 1,4-dimethylcyclohexanes; b) cis and trans 1,3-dimethylcyclohexanes; c) mixtures of isomers of these compounds.

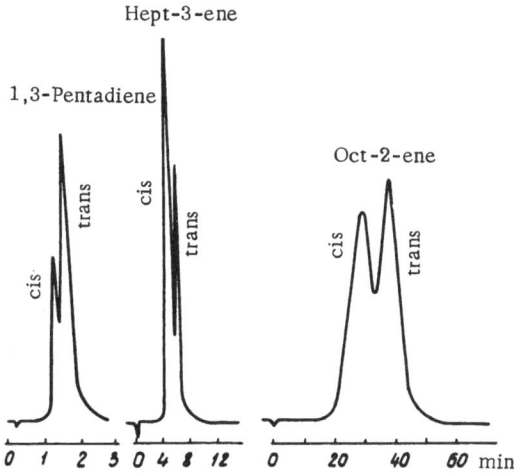

Fig. 16. Mixture of isomers of 1,3-pentadiene, hept-3-ene, and oct-2-ene, conditions as in Fig. 15.

Table 3. Number of Carbon Atoms n, Molecular Weight M, Boiling Point, Retained Volume V_s, and Isosteric Heat of Adsorption Q

No.	Compound	n	M	B.p.	V_s, ml/m²					Q kcal/mole
					100°	125°	150°	200°	230°	
1	Pent-2-ene (cis)	5	70	36.94	0.63	—	—	—	—	7.6
2	Pent-2-ene (trans)	5	70	36.35	0.63	—	—	—	—	8.7
3	1,3-Pentadiene (cis)	5	68	44.07	1.05	0.52	0.23	0.13	0.07	8.4
4	1,3-Pentadiene (trans)	5	68	42.03	1.29	0.64	0.29	0.13	0.07	9.1
5	Hex-2-ene (cis)	6	84	68.84	2.69	1.14	—	—	—	8.9
6	Hex-2-ene (trans)	6	84	67.88	3.34	1.34	—	—	—	10.0
7	Hept-2-ene (cis)	7	98	98.50	10.12	4.12	1.90	0.90	0.64	10.3
8	Hept-2-ene (trans)	7	98	97.80	12.15	4.72	1.96	0.90	0.64	11.4
9	Hept-3-ene (cis)	7	98	95.80	8.20	4.02	2.36	0.67	—	9.1
10	Hept-3-ene (trans)	7	98	95.70	10.26	4.71	2.36	0.67	—	10.7
11	1,4-Dimethylcyclohexane (cis)	8	112	124	7.13	3.67	1.64	0.43	0.25	10.6
12	1,3-Dimethylcyclohexane (trans)	8	112	124	8.25	3.92	1.74	0.47	0.27	11.0
13	1,3-Dimethylcyclohexane (cis)	8	112	120	14.34	6.80	2.81	0.73	0.30	11.4
14	1,4-Dimethylcyclohexane (trans)	8	112	119	20.0	8.11	3.36	0.80	0.41	11.9
15	Oct-2-ene (cis)	8	112	123	30.0	12.30	4.90	1.24	—	11.6
16	Oct-2-ene (trans)	8	112	123	33.80	13.65	5.30	1.24	—	12.7

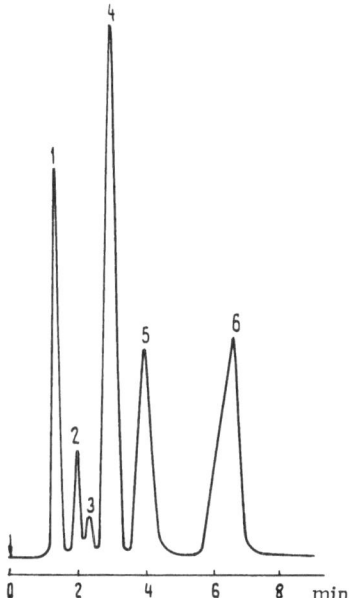

Fig. 17. Stereoisomers on a 100 × 0.4 cm column of graphitized carbon black (s = 8.1 m²/g) at 200°: 1) trans-1,2-bis(trimethylsilyl)ethylene; 2) cis-1,2-bis(trimethylsilyl)ethylene; 3) trans-1-trimethylgermyl-2-trimethylsilylethylene; 4) cis-1-trimethylgermyl-2-trimethylsilylethylene; 5) trans-1,2-bis(trimethylgermyl)-ethylene; 6) cis-1,2-bis(trimethylgermyl)ethylene.

but-2-ene (b.p. 0.88°C) it is 0.194. Even small differences in boiling point, as for the pent-2-enes, produce appreciably different times of emergence; V_m for cis-pent-2-ene is 0.81, while that for trans-pent-2-ene is 0.92. The two hex-2-enes have the same boiling point (67.9°C), but the cis compound has a V_m of 2.31, while the trans one has 2.69. For the hept-3-enes, which both boil at 95.8°C, the values are 6.03 (cis) and 6.92 (trans).

It has been shown [148] that the cis and trans isomers of the olefins show a linear relation of log V_m to boiling point, the line for the cis compounds running below that for the trans ones, on account of difference in adsorption energy, which is governed by the orientation on the surface, that of the trans forms being more favorable on graphitized carbon black. The chromatograms (Fig. 16) show that graphitized carbon black separates the cis and trans isomers of the following: 1,3-pentadiene, hept-3-ene, and oct-2-ene. The cis isomer emerges first.

The trans isomers emerge before the cis ones in the case of 1,2-bis(trimethylsilyl)ethylenes and their germanium analogs (Fig. 17), since the cis forms have more links in contact with the surface.

Separation of the terpenes [37] is another example where the dominant factor is the geometrical structure, with very little effect from the groups A, B, and D. Figure 18 shows results for terpenes and n-alkanes (the latter used as standard representatives of group

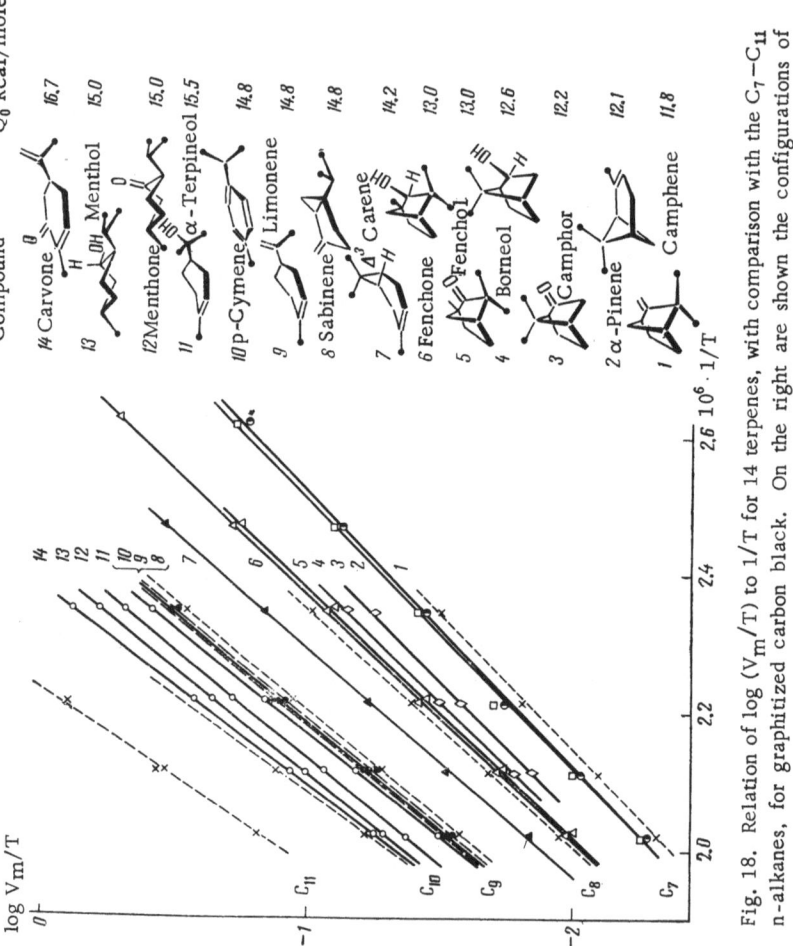

Fig. 18. Relation of log (V_m/T) to 1/T for 14 terpenes, with comparison with the C_7–C_{11} n-alkanes, for graphitized carbon black. On the right are shown the configurations of the terpene molecules and the Q_0.

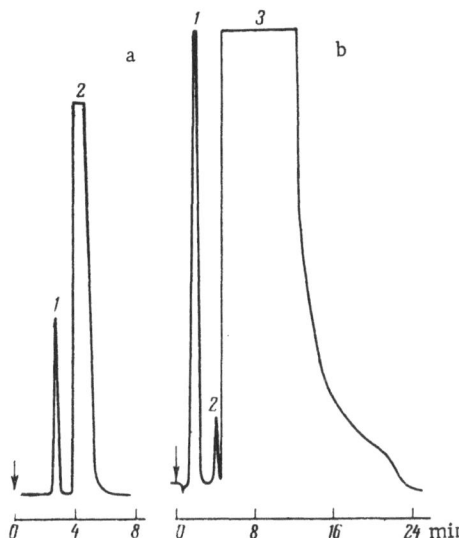

Fig. 19. a) Mixture of thiophen (1) and benzene
(2) run at 100°C on a 200 × 0.5 cm column of
graphitized carbon black (s = 7.6 m²/g, helium
50 ml/min, katharometer); b) mixture of methane (1), acetylene (2), and ethylene (3) on a
100 × 0.4 cm column (s = 70 m²/g, nitrogen 50
ml/min, flame ionization.

A) in terms of log (V_m/T) against $1/T$ (T is the absolute temperature of the column), with the Q_0 deduced from these. The largest
Q_0 and V_m occur for the most nearly planar molecules; these quantities are directly related to the number of C and O atoms contacting the surface. A convenient parameter here, which indicates the
order of emergence from the column, is the sum of the reciprocals
of the distances of the C and O atoms from the surface in the most
favorable orientation [37].

Isolation of Trace Components. Columns of graphitized carbon black isolate certain impurities ahead of the main
peak, which makes analysis for these particularly accurate. Figure
19a shows the thiophen peak considerably ahead of the peak for benzene (main component) [29, 34]; similarly, Fig. 19b shows the acetylene peak ahead of the main ethylene peak [38].

   Theoretical Calculation of V_m. The simple lattice structure of graphite and the nonspecific adsorption have led
[31, 33, 39-61] to detailed studies, such as calculation of the potential energy Φ and statistical derivation of the thermodynamic
parameters. The nonspecific adsorption on a uniform surface allows Φ to be calculated as a function of the coordinates of the center
of mass and the orientation relative to the basal plane. The statistical theory also gives satisfactory agreement with experiment
for the change in the chemical potential, which is related to the
Henry constant, as well as ΔS_0 and Q_0.

   V_s for small concentrations is the Henry constant, so this
may also be derived theoretically. Poshkus [59, 60] has derived
expressions for the V_s of a complex mixture at low loadings. In
the quasiclassical approximation for rigid polyatomic molecules

$$V_s \approx \frac{1}{8\pi^2 ms} \int \cdots \int [\exp(-\Phi/kT) - 1] \sin \vartheta \, dx dy dz d\vartheta d\varphi d\psi, \quad \text{(II, 3)}$$

in which m and s are the mass and specific surface of the adsorbent; x, y, z are the Cartesian coordinates of the center of gravity
of the molecule; and ϑ, φ, ψ are the normal Euler angles defining
the orientation relative to the surface, ϑ being the angle between
the principal axis of the molecule and the z axis, which is perpendicular to the surface. Only the Cartesian coordinates of the center
persist for a monatomic substance.

   A Buckingham potential (two power terms for attraction and
an exponential term for repulsion) was used for the interaction potential of the force centers with the semiinfinite lattice; the constant for the dispersion attraction was deduced from the Kirkwood–
Müller formula [62, 63] for the first term and from an analogous
formula [64] for the second in terms of the polarizability and dia-
magnetic susceptibility for the carbon atoms of the adsorbent and
for the atoms or groups in the molecule. The exponential repulsion constant was derived as the geometric mean of these constants
for adsorbent and adsorbate, while the preexponential repulsion factor was deduced from the equilibrium condition (i.e., minimum Φ
at a distance determined in terms of the van der Waals radii for
the molecule and the interplanar distance for graphite). The interaction of all parts of the molecule with the semiinfinite lattice was
introduced by summation over the atoms [39, 48, 65] or by integration within planes and summation with respect to planes [42, 58-60].

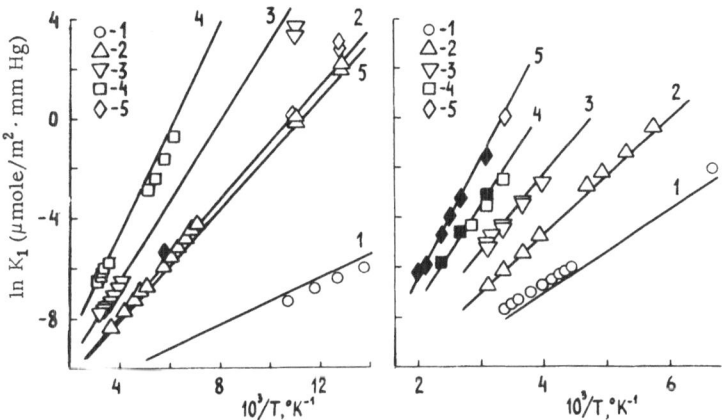

Fig. 20. Relation of $\ln K_1$ (Henry constant) to $1/T$ for various substances. Left: 1) Ne; 2) Ar; 3) Kr; 4) Xe; 5) N_2. Right: 1) methane; 2) ethane; 3) propane; 4) n-butane; 5) n-pentane. The lines are from theory for adsorption on the basal face of a semiinfinite graphite lattice; the points are from experiment for graphitized thermal carbon black.

Figure 20 shows theoretical results for log V_s as a function of T for various substances on basal planes of graphite (these calculations involved various simplifying assumptions on the motion of the molecules in the adsorbed state) [59, 60]; these are compared with observed results for graphitized thermal carbon black as deduced from adsorption or gas-chromatographic measurements (see [59, 60] for references to the experimental data). The theoretical results agree satisfactorily with experiment, which shows that the nonspecific interaction is related in the main to dispersion forces. Calculated V_s for the lower alkanes on graphite [66] are also in good agreement with experiment. Here semiempirical potential functions was used for the interaction of graphite C with C and H atoms, and also CH_3 and CH_2 groups, in the molecule [66, 150]. Allowance has been made [150, 151] for the adsorption of the various rotational isomers for n-butane and n-pentane.

Graphite is also of interest in relation to molecules with conjugated bonds, such as butadiene, polyacenes, tetracyanoquinodimethane, etc. However, there are only small effects from conjugation in adsorption on nonspecific adsorbents, in particular the basal plane of graphite, as may be seen [35] for separations on graphitized thermal carbon black. This material is of considerable interest as regards the separation of compounds containing H and D.

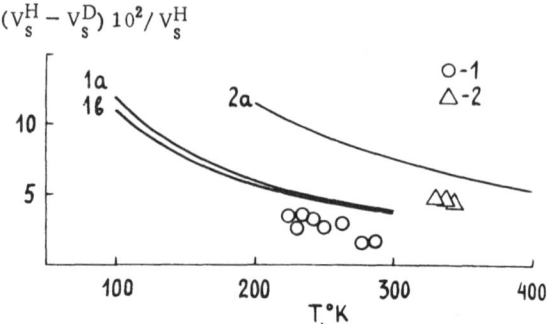

Fig. 21. Calculated curves and observed values (points) for $R = (V_s^H - V_s^D)/V_s^H$ graphite (curves) and for graphitized thermal carbon black (points) for CH_4, CD_4, C_6H_6, and C_6D_6 [155, 156]. Curve 1a was calculated without the correction for quantization of the vibration of the center of mass perpendicular to the surface; curve 1b was calculated with that correction.

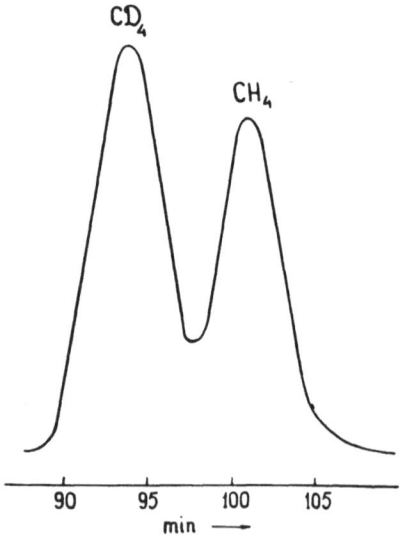

Fig. 22. Mixture of CD_4 and CH_4 on Saran charcoal at 0°C, column 5 m long, 2 mm in diameter, helium 43 cm³/min, flame-ionization detector.

Kiselev and Poshkus [152, 153] have calculated $R = (V_s^H - V_s^D)/V_s^H$
theoretically, and also $Q^H - Q^D$ for CH_4, CD_4, C_2H_6, C_2D_2, C_3H_8,
C_3D_8, C_6H_6, and C_6D_6 on the basal face of graphite. These differences are governed mainly by the difference in the potential energies of adsorption; replacement of H by D reduces the statistical
mean polarizability of the molecule. Figure 21 shows R as calculated (curves) [154] and as found by experiment (points) [155] for
CH_4, CD_4, C_6H_6, and C_6D_6. The theory predicts the separation correctly. Figure 22 shows that the sequence predicted theoretically
for CH_4 and CD_4 on graphite is actually observed [156, 157] for
Saran charcoal.

Avgul' and Kiselev [157] have considered in detail the molecular theory of adsorption on graphite and the experimental evidence for graphitized carbon black.

Other Nonspecific Adsorbents.   Certain other adsorbents with layer lattices (BN and MoS_2) are also nonspecific, but
it is difficult [67] to make these in homogeneous form. Moreover,
MoS_2 is readily oxidized.

Nonspecifically adsorbing surfaces may be produced by chemical modification, e.g., of macroporous silica gel by attachment of
trimethylsilyl groups by reaction with trimethylchlorosilane [68-
73] or hexamethyldisilazane [72, 74, 75]. Attachment of a close-
packed layer of these groups is difficult on account of steric hindrance during reaction of the hydroxylated surface with the first
reagent [69, 70, 76], but layers sufficiently dense to prevent the
penetration of large hydrocarbon molecules have been produced
[69, 70, 77]. The van der Waals distances between the methyl groups
in the trimethylsilyl coating are much greater than the bond lengths,
so the concentration of force centers is much less than that for the
unmodified surface, and hence the energy of adsorption for a hydrocarbon is very small (less than the heat of condensation [70, 78]);
even the room-temperature adsorption isotherms for n-hexane
and benzene are initially linear. Heavy hydrocarbons can be separated with such surfaces. For example, hexadecane comes off a
50 cm column of modified wide-pore silica gel in 1.5 min at 200°C.
The surface of the silica must be specially prepared [73] for reaction with trimethylchlorosilane.

Similar results are obtained by attaching alkoxy groups by
reaction of alcohols with strongly dehydroxylated silica [79-83].

TYPE II ADSORBENTS

These bear localized positive charges (acid centers or exchangeable cations), as in hydroxylated silica and cationized zeolites.

Hydroxylated Silica Gel: Effects of Dehydroxylation. The specific molecular interaction of a surface functional group is dependent on the skeletal atom bearing it; OH linked to silicon acts as a weak acid, with the H partly protonized [11, 84]. The surface concentration of the OH groups is also important; special methods of determination are needed here.

The most convenient and reliable method is exchange of OH with D_2O or OD with H_2O [85-89], which rapidly goes to completion. Figure 23 shows α_{OH} as a function of T in vacuum treatment (up to 900-950°C) as deduced via a range of surface reactions [86-88, 90] for nonporous and macroporous specimens with previously fully hydroxylated surfaces. Here T is the decisive factor, the specific surface having hardly any effect [87, 88, 91] if the surface is fully hydroxylated. The α_{OH} after prolonged pumping at 150-200°C is 8-9 μmole/m^2 or about five OH per 100 Å^2. The curve of Fig. 23 enables one to produce any desired α_{OH}. Marked sintering below 900-950°C occurs only for gels with relatively narrow pores [92-94].

IR spectra of silica gel (not ultraporous) after heat treatment show that prolonged evacuation at 150-200°C removes virtually all the adsorbed H_2O (the absorption in the deformation bands of H_2O vanishes [87, 89, 95, 96]). The hydroxylated surface at up to 350-400° releases mainly hydrogen-bonded OH groups, which leaves only free groups, which give a narrow band at ~3750 cm^{-1} (OH) or ~2760 cm^{-1} (OD) [87, 89, 95-99]. These free groups are slowly lost at higher temperatures.

Silica gel also usually contains OH groups within the particles, the absorption peak for these being at ~3650 cm^{-1} [87, 89, 99]. The heating loss usually fails to give α_{OH} correctly, since these internal groups are also lost [87, 88, 91, 99]. The content of internal groups may be deduced from the difference between twice the water loss on firing and the content of surface OH groups. This internal content is dependent on the heat treatment and the particle size, i.e., on the specific surface [87, 91, 99].

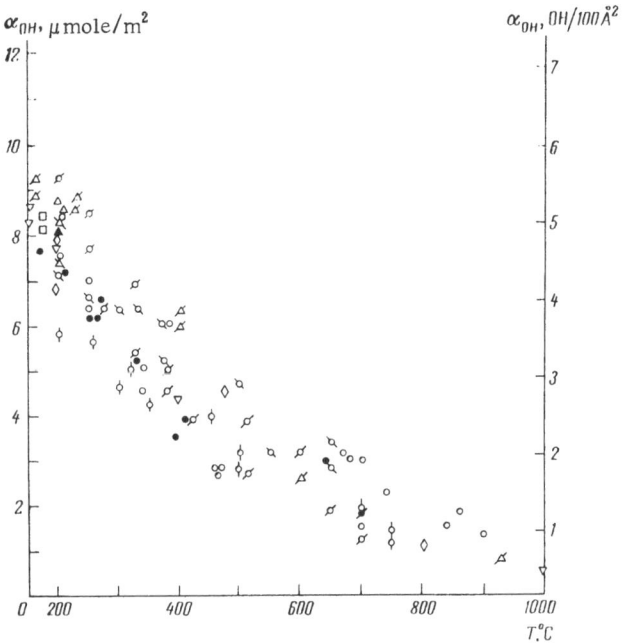

Fig. 23. Mean OH concentration on silica as a function of
treatment temperature (initial state maximally hydroxylated). The various points relate to specific surfaces ranging from 39 to 750 m^2/g and to reactions with D_2O, CH_3Li,
and CH_3MgI.

It is convenient to compare the maximally hydroxylated state
(after evacuation at ~200°C) with the maximally dehydroxylated
state (after heating to 900-950°C, which produces very little sintering) in order to elucidate the effects of α_{OH} on the behavior for
nonporous or wide-pore silica gel. Figure 23 shows that the surface after the latter treatment retains about 10% of the initial OH
groups.

The involvement of surface OH in the molecular interaction
can be deduced from IR spectra and from the adsorption behavior,
both as examined before and after dehydroxylation.

Figure 24 [3-5, 100] shows IR spectra for Aerosil of specific
surface ~180 m^2/g following evacuation at 400°C after full hydroxylation, and also spectra on adsorption of increasing quantities of substances of group B (ones that interact specifically with protonized

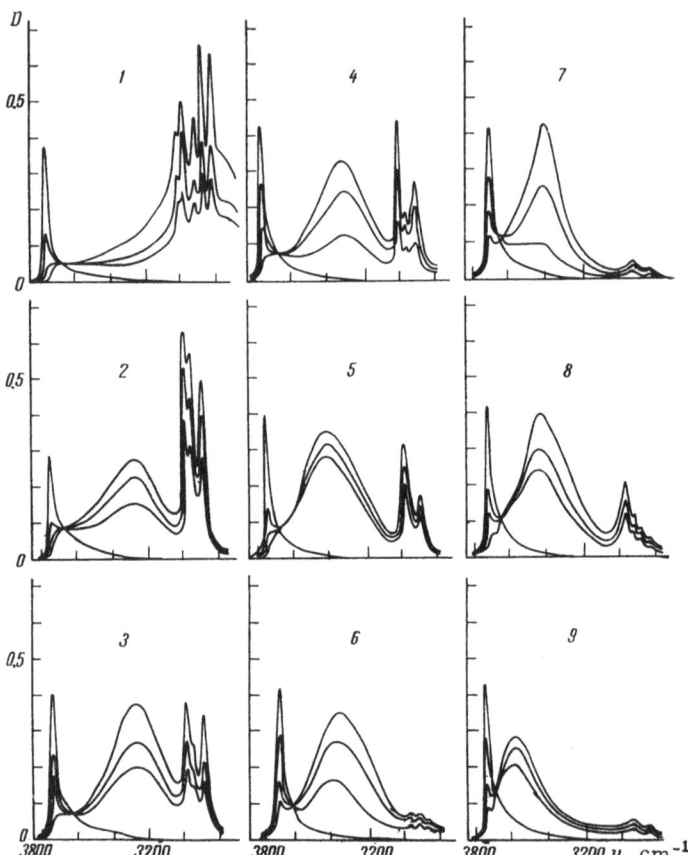

Fig. 24. Changes in IR spectra of Aerosil (evacuated at 450°C) in the region of the OH stretching frequency produced by increasing quantities of group B compounds: 1) trimethylamine; 2) dibutyl ether; 3) tetrahydrofuran; 4) diethyl ether; 5) cyclopentanone; 6) acetone; 7) acetonitrile; 8) ethyl acetate; 9) nitromethane. D is optical density.

H atoms of surface OH groups, Table 1).* The height of the 3750 cm^{-1} peak (free OH)† is reduced, while a broader band due to OH groups perturbed by adsorption appears at lower frequencies.‡

---

* The strength of the ~ 3650 cm^{-1} band from internal OH (unaffected by adsorption) is low in these specimens.

† The bands at 2800-3000 cm^{-1} are CH modes of the adsorbed compounds.

‡ In the case of trimethylamine, the perturbed OH band overlaps the CH bands of the compound.

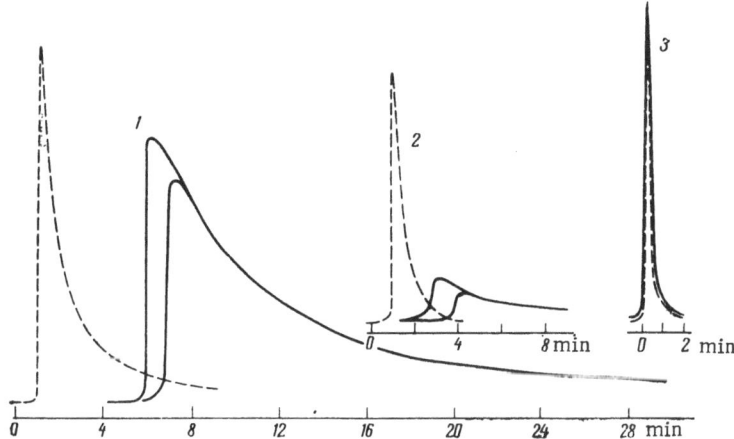

Fig. 25. Chromatograms of: 1) diethyl ether; 2) mesitylene; 3) n-dodecane on hydroxylated (full lines) and greatly dehydroxylated (broken lines) macroporous silica gel at 200°C (column 20 × 0.5 cm, nitrogen, flame ionization).

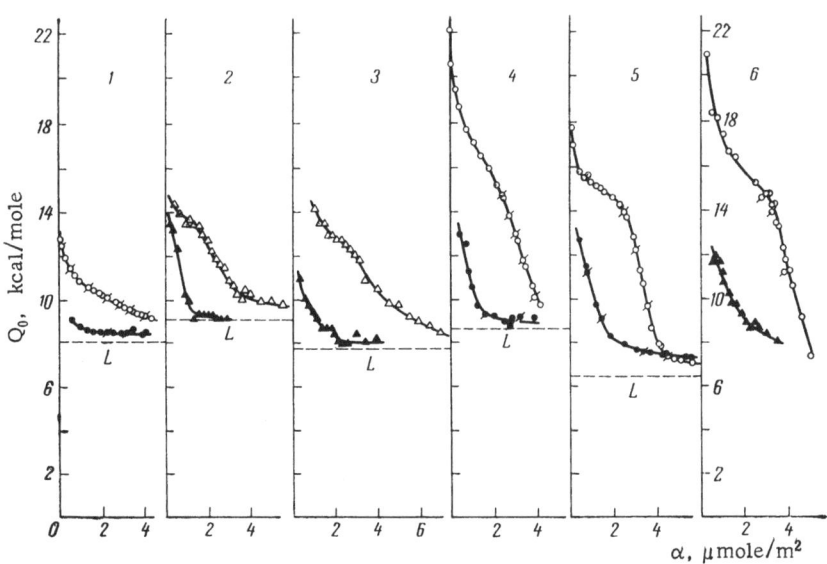

Fig. 26. Relation of Q_0 for group B compounds to α for hydroxylated (open symbols) and greatly dehydroxylated (filled symbols) macroporous silica gel: 1) benzene; 2) nitromethane; 3) acetonitrile; 4) ethyl acetate; 5) diethyl ether; 6) tetrahydrofuran. L is the latent heat of condensation.

The shift $\Delta\nu_{OH}$ and the intensity change are very much dependent on the electronic structure [1-5, 95, 96, 100-103]. Saturated hydrocarbons (group A) produce only very small $\Delta\nu_{OH}$, while an aromatic hydrocarbon with the same number of C atoms produces a somewhat larger shift. The shifts produced by polar functional groups increase from NO_2 to CN, CO, ester, and amine. The $N\overset{\prime}{O}_2$ group in nitromethane produces only a relatively small shift, in spite of the very high dipole moment (several times that of an ester or amine), which means that these specific interactions cannot be described in terms of the dipole moment alone, but are due rather to peripheral details of the electron density. A given functional group produces a largely constant $\Delta\nu_{OH}$ when it appears in different molecules, e.g., aliphatic and cyclic ethers (diethyl and dibutyl ethers, tetrahydrofuran) or ketones (acetone and cyclopentanone).

Similar changes in the stretching frequency occur for OH in alcohols [104] and CH in alkynes [105] when these interact in solution with molecules of groups A and B; hence the nature of the interaction is as for the surface case. This facilitates comparison of gas − liquid and gas −adsorption types of chromatography.

The shift and intensity change do not give a direct indication of the interaction in energy terms, but they do arise solely from interaction with the OH groups, so $\Delta\nu_{OH}$ and the intensity change should be compared not with the total adsorption energy but with the part due to the specific interaction with OH as determined independently [5, 106].

Group A molecules (inert gases, CCl_4, saturated hydrocarbons) give Q_0 almost independent of the degree of hydroxylation for wide-pore silica gel, so the strongly protonized H atoms have no appreciable effect on this energy, the interaction remaining largely unspecific. The isotherms for the surface concentration α (uptake per unit surface) are thus nearly the same in all states for this group [107-110]. Figure 25 shows (for macroporous silica gel of only ~20 m²/g, dehydroxylated at 350 and 1000°C) that the peaks for saturated hydrocarbons are symmetrical and almost coincide [3, 4, 111] in spite of the great change in α_{OH}.

Group B gives an entirely different picture, since such compounds interact specifically with partly protonized H; a difference from group A is that dehydroxylation reduces Q substantially [1-5,

100, 107, 109, 110, 112] (Fig. 26) and also the absolute uptake per unit area, as well as the delay at a given concentration (peak height) as Fig. 25 shows for mesitylene and diethyl ether. Silica gel often contains traces of alumina, which produces not only Brönsted acid centers (OH groups) but also strong Lewis acid centers. Aerosilogels [158-160] are of high chemical purity and high thermal stability; they often give highly symmetrical peaks [159, 160]. The asymmetrical peaks in Fig. 25 for group B molecules are due to these admixtures.

A strong specific interaction with protonized H occurs when there is a bond or group with peripherally localized electron density. The difference of the Q for hydroxylated and dehydroxylated surfaces may be taken as a measure of the contribution from the specific OH interaction to the total interaction energy. It is somewhat difficult to measure this difference ΔQ because it is dependent on θ, since silica gel has an inhomogeneous surface. The difference is less for θ small, since even gel dehydroxylated at 900°C retains some 10% of its hydroxyl groups (Fig. 23). Again, ΔQ decreases at high θ because we get polymolecular adsorption, and Q tends to L (heat of condensation) for any surface. Hence the contribution from the specific interaction may be taken as ΔQ for $\theta \approx 0.5$ [1, 3-5, 100, 109, 110, 112]:

$$\Delta Q_{(\theta \approx 0.5)} = Q_{h \cdot (\theta \approx 0.5)} - Q_{d \, (\theta \approx 0.5)}$$

(II, 4)

Dehydroxylation causes Q for diethyl ether (group B) to approach Q for n-pentane (a corresponding compound of group A), as for unspecific adsorbents of type I (Table 2). Also, $Q_{d(\theta \approx 0.5)}$ for group B is* close to L, so

$$\Delta Q_{(\theta \approx 0.5)} \simeq Q_{h \cdot (\theta \approx 0.5)} - L.$$

(II, 5)

Figure 27 [100, 103] shows that these ΔQ are correlated with the $\Delta \nu_{OH}$ [3, 4, 100, 102, 103] and with the intensity changes [4, 106]. Now $\Delta Q \approx 0$ for group A, and the $\Delta \nu_{OH}$ are small, so the re-

---

*The Q for saturated hydrocarbons at $\theta \approx 0.5$ for nonporous and wide-pore silica gels (hydroxylated or otherwise) are also close to the corresponding L [107, 110, 113].

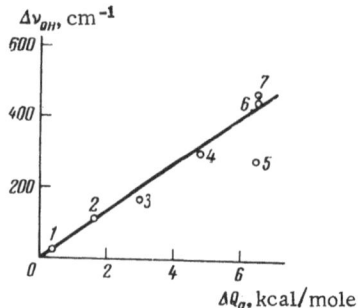

$\Delta\nu_{OH}$, cm^{-1}

Fig. 27. Relation of $\Delta\nu_{OH}$ to ΔQ (as measured for hydroxylated and largely dehydroxylated surfaces) for: 1) n-hexane; 2) benzene; 3) nitromethane; 4) acetonitrile; 5) ethyl acetate; 6) diethyl ether; 7) tetrahydrofuran.

lation of $\Delta\nu_{OH}$ to $\Delta Q_{(\theta \approx 0.5)}$ is nearly one of direct proportion for compounds containing oxygen.*

ΔQ and $\Delta\nu_{OH}$ also have a fairly simple relation to V_S for hydroxylated silica gel [3-5], so the adsorption on these surfaces can be correlated with spectroscopic and energy features of the specific interaction. For group A (e.g., n-dodecane), $V_{s.h}/V_{s.d} \approx 1$, so this group is largely unaffected by surface OH, whereas group B shows a marked effect. The $V_{s.h}/V_{s.d}$ for compounds in group B with the same functional group are similar, e.g., for diethyl ether and tetrahydrofuran [111].

These specific-interaction effects are related not so much to the dipole moment as to more detailed aspects of the peripheral electron density. Nitrogen, ethylene, and benzene are nonpolar, but they show a fairly strong specific interaction with OH, whereas nitromethane (very high dipole moment) shows only a relatively weak specific interaction. A hydrogen bond may [1-5] be considered as a particular case of specific interaction, so we may expect generally similar effects whenever a group B compound (one with peripherally localized electron density) interacts with a surface having positive charge localized on particles of small radius. In fact, similar effects occur not only for acidic hydroxylated surfaces but also for surfaces bearing cations.

Cationized Zeolite Surfaces. The walls in the channels of porous zeolites consist of Si—O and Al—O tetrahedra [12-14]. The coordination number of Al (four) exceeds the valency (three), so the Al—O tetrahedra have excess negative charge, which is distributed over the numerous Al—O—Si bonds within the framework, whereas the compensating positive charge is localized on exchangeable cations, usually of small radius. This charge distri-

* The $\Delta\nu_{OH}$ are averaged, whereas the ΔQ are additive, if the molecule is polyfunctional and all functional groups are in contact with the surface.

Table 4. Zeolite Q_0 Derived from Chromatograms

Zeolite	Q_u, kcal/mole								$Q_{C_2H_4} - Q_{C_2H_6}$	$Q_{C_3H_6} - Q_{C_3H_8}$
	O_2	N_2	CO	CH_4	C_2H_6	C_2H_4	C_3H_8	C_3H_6		
NaX, crystals	—	—	—	4.5	6.0	9.1	7.9	—	3.1	—
NaX with binder	3.1	5.3	—	—	—	8.8	8.1	—	—	—
CaX with binder	4.1	5.7	7.2	4.7	7.8	12.6	8.8	—	4.8	—
CaA with binder	4.3	6.3	7.5	5.8	8.0	12.8	9.3	13.7	4.8	4.4

bution is somewhat similar to that on a hydroxylated acidic oxide, e.g., silica, so we expect generally similar behavior. In fact, zeolites retain nitrogen much more strongly than argon or oxygen, and ethylene much more strongly than ethane. Table 4 shows that Q_0 for group B (e.g., ethylene, propene) are much larger than those for the corresponding group A compounds (ethane, propane); the differences between corresponding pairs reflect the contribution from the specific interaction. Table 4 shows that this contribution increases rapidly with the charge of the compensating cation [3, 114].

The surfaces are fairly homogeneous as regards group A compounds, especially for LiX, since the small Li^+ cations readily enter the gaps between O atoms in the framework, and, because of their low polarizability, make little contribution to the dispersion potential. LiX then clearly reveals the adsorbate–adsorbate interaction; the isotherms and Q as functions of θ resemble those for graphitized carbon black in the case of group A. Xe [25, 115] and propane [116] give Q increasing with θ for LiX, while the adsorption isotherms are initially convex to the pressure axis. Figure 28 shows such curves for Xe on LiX and NaX; as for grapitized carbon (Fig. 2), the results are described satisfactorily by (II, 1) as functions of p and T. The isotherms of Fig. 28 (unspecific adsorption on porous zeolite crystals) are also well described by (II, 1a), while (II, 1c) describes the dependence of the uptake on p and T [147]. These equations describe closely the nonspecific adsorption of hydrocarbons by zeolites, and also the specific adsorption of CO_2, zeolites having inhomogeneous surfaces in relation to this [145-147] (in this case, the heat of adsorption decreases as θ increases). In accordance with (II, 1) and (II, 1c), at the low θ and

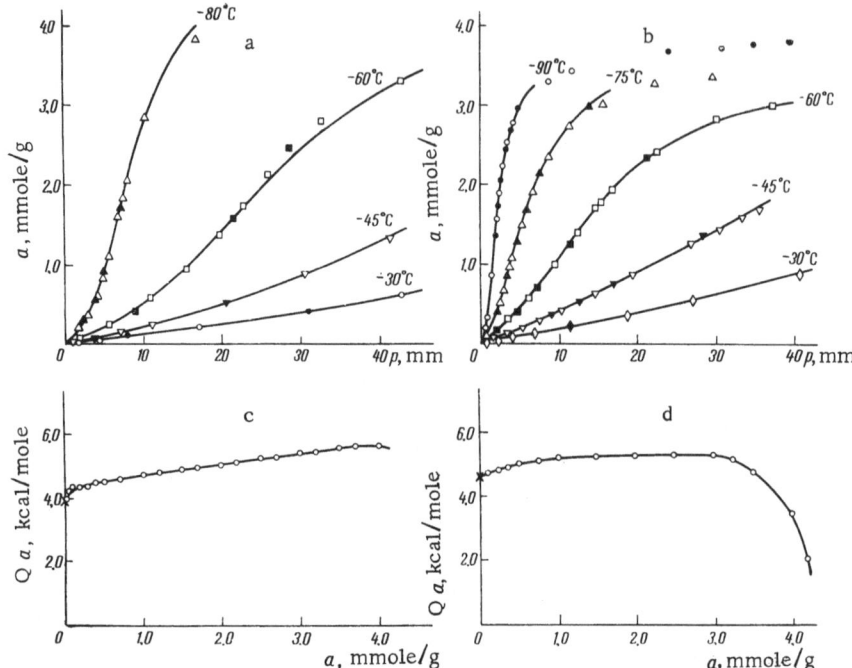

Fig. 28. Adsorption isotherms for Xe for a) LiX and b) NaX zeolites at various temperatures, and relation of Q_a for Xe to uptake a for c) LiX and d) NaX. Curves are calculated from Eq. (II, 1). Point symbols are experimental data. Filled symbols indicate desorption.

high T characteristic of chromatography columns there is a clear-cut Henry's-law region, and the peaks are therefore symmetrical.

Figure 29 illustrates the effects of cation radius (for fixed charge) on the interactions with compounds of groups A and B by reference to Q_a for n-pentane (group A, nonspecific) and diethyl ether (group B, specific and unspecific) on X zeolite bearing Li$^+$, Na$^+$, K$^+$, Rb$^+$, and Cs$^+$ [3, 4, 117, 118]. Rb$^+$ and Cs$^+$ (large) project far into the channels [119], and they also have high polarizability (from their many electrons), so they show high energies of non-specific interaction; the Q_a for the alkane increase along this series. The total number of alkane molecules that can be accommodated in the channels decreases along the series, especially for RbX and CsX.

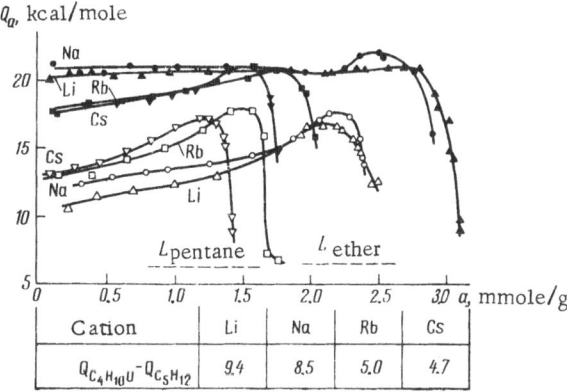

Fig. 29. a) Relation of Q_a for diethyl ether (filled symbols) and n-pentane (open symbols) to uptake a for LiX, NaX, RbX, and CsX; b) ΔQ (between compounds) in relation to cation.

The ether shows considerable specific interaction, mainly from the unpaired electrons of the O to the cations (the interaction is probably of quantum-mechanical type at these high energies), but with some interaction of the ether dipoles with the fields set up by the cations and negative charges in the zeolite lattice.

Dimethyl ether and propane have nearly equal L and similar Q for nonspecific adsorption on graphitized carbon and dehydroxylated silica gel, as do diethyl ether and n-pentane (Table 2) [1-5, 112]. Hence the ΔQ_0 as between the first pair [116] or the second [117] at low θ for X zeolite,

$$\Delta Q_0 = Q_{0(CH_3)_2O} - Q_{0(CH_3)_2CH_2} \approx Q_{0(C_2H_5)_2O} - Q_{0(C_2H_5)CH_2} \qquad (II, 6)$$

are very large, especially for Li^+ and Na^+. This ΔQ_0 is thus a rough measure of the energy of specific interaction for the oxygen.

The energy of the nonspecific interaction for the ether increases with cation radius roughly as for the n-alkane, whereas the specific interaction tends to fall, and so ΔQ for diethyl ether and n-pentane (Fig. 29) is much the same for LiX as for NaX (the two change by roughly equal amounts), whereas there is a large fall to RbX and CsX (nearly to half for CsX) [3-5, 117, 118].

It is thus possible to adjust the energy and specificity of the interaction of group B compounds within wide limits by choice of

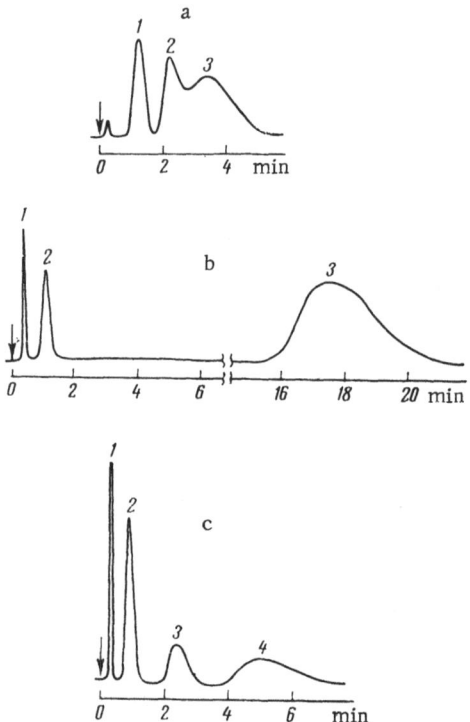

Fig. 30. Chromatograms on a 50 × 0.5 cm CaA
column: a) 240℃: 1) cyclohexane; 2) ethylene;
3) propane; b) 300℃: 1) cyclohexane; 2) benzene; 3) n-hexane; c) 300℃: 1) cyclohexane;
2) benzene; 3) toluene; 4) ethylbenzene.

charge, radius and concentration (which partly involves alteration
of the skeletal Al: Si ratio) of exchange cations.

Aromatic hydrocarbons, and group B compounds having polar groups (CH_3NO_2, CH_3CN), are so strongly adsorbed in channels
of sizes that will admit the molecules (the X zeolites) that gas-
chromatographic separation of them is virtually impossible, even
at very high temperatures. However, the adsorption may be studied
by use of the more weakly adsorbing outer surface of the porous
crystals; adsorption on the internal surface may be avoided by use
of a zeolite whose channels are too small to admit the molecules.
For instance, NaA zeolite has very small pores, and the outer surface of these crystals provides sharper and more rapid separation

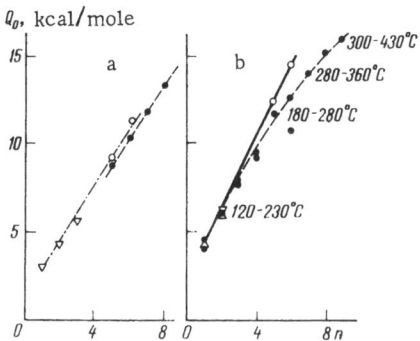

Fig. 31. Q_0 for n-alkanes as a function of chain length n for: a) graphitized carbon black; b) NaX crystals as determined statically (O calori-metrically, ∇ isosterically) and by gas chromatography ($\bullet$).

Table 5. Retention Relative to n–Heptane at 100°C for Compounds Supported on Macroporous Alumina

Compound	Cyclohexane	Isooctane	Hept-1-ene	Benzene
NaOH	0.36	1.21	1.34	1.05
NaCl	0.28	1.28	1.80	1.97
NaBr	0.27	1.32	2.00	3.00
NaI	0.30	1.37	2.31	3.90

of aromatic hydrocarbons than does the internal surface of NaX [3, 4, 120]. Curves for the outer surface of CaA (Fig. 30) and the in-ternal surface of NaX show rapid equilibration in the first case, but this occurs only for groups A (weakly adsorbed) in the second case, and that for small molecules, as is clear from Fig. 31, which gives Q_0 derived from static and gas–chromatographic tests for n-alkanes on nonporous graphitized carbon black [28, 32, 121-123] and porous NaX zeolite [4, 124-128]. The peaks given by NaX be-come broader as the chain lengthens, on account of slow exchange in the fine pores. Some polar compounds of group B can be sepa-rated on the outer surfaces of NaA and CaA zeolites. Various por-ous crystals have been used in gas chromatography [129].

Salts. Nonporous crystals of simple structure (salts, especially alkali halides) are of considerable interest in gas chromatography. Scott [15] used salts with a given cation having anions of different sizes: NaCl, NaBr, NaI, and also NaOH, these being deposited in large quantities on macroporous alumina. The H in NaOH is not protonized, but the entire OH group acts as an anion of small radius. Table 5 shows that the specific interaction with group B compounds having π-bonds is in inverse relation to the size of the anion. Good results in the separation of complex mixtures have been obtained with salts [130]; the uses of various salts are described in [131], and also in other papers dealt with in Chapter V.

TYPE III ADSORBENTS

This type (Table 1) has localized negative charges, such as small anions (if the corresponding cations are large) or the anions in the outer layer of a platy crystal (e.g., $NiCl_2$), as well as functional groups with atoms bearing nonbonding pairs on the outside.

It is difficult to use anion-exchange materials as type III adsorbents, because they often have other functional groups, e.g., OH. Salts resembling $NiCl_2$ have also not yet been used in gas chromatography, although they may have very homogeneous surfaces. It is convenient to produce peripherally localized electron density via functional groups of group B type, e.g., oxygen in ether or carbonyl (but not OH) groups, nitrogen in cyanide or tertiary amine groups (but not secondary or primary ones), etc. These groups may be emplaced chemically or by adsorption modification.

For example, macroporous silica gel has [4] been chemically modified by the attachment of organosilicon groups bearing CN at the end. The material then binds strongly compounds of group D, which have hydrogen atoms in alcohol and amine groups, as Fig. 32a shows by reference to log V_s for n-alkanes, n-alkylbenzenes, ethers, n-alkyl cyanides, nitro derivatives of n-alkanes, and n-alcohols. The surface CN forms a hydrogen bond to OH, so alcohols (group D) are held strongly; moreover, CN has a large dipole moment, so other compounds not capable of hydrogen bonding are also firmly held if they have high dipole moments, such as cyanides and nitro derivatives (group B). Also, group C compounds should be strongly bound.

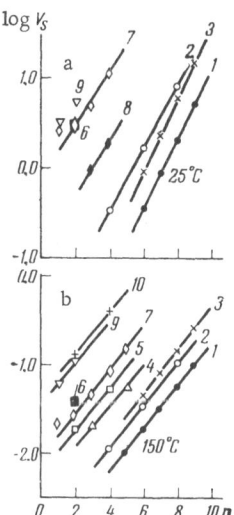

Fig. 32. Relation of log V_s to number of carbon atoms n for: a) macroporous silica gel chemically modified by organosilicon groups terminating in CN; b) graphitized channel black bearing a close-packed monolayer of polyethylene glycol: 1) n-alkanes; 2) ethers; 3) n-alkylbenzenes; 4) ketones; 5) n-alkyl bromides; 6) n-alkyl cyanides; 7) n-alcohols; 8) esters; 9) n-nitroalkanes; 10) n-amines.

Chemical modification gives adsorbents with adequate thermal stability having surfaces of a great variety of compositions and hence varying in selectivity; but it is difficult to produce a sufficient density of attached groups on account of steric hindrance [69, 70, 132]. Moreover, the base materials are usually oxides or hydroxides, which have geometrically inhomogeneous surfaces. Therefore adsorption modification is of considerable interest, as close-packed monolayers can be used, and these differ from the relatively thick films of liquid commonly used in gas-liquid chromatography in that they lie within the molecular field of the base, so the volatility is much reduced relative to the bulk liquid, which is of particular value at high column temperatures.

The force centers in adjacent molecules of such a monolayer are separated by van der Waals distances, which substantially exceed the distances between chemically bonded pairs of atoms in the molecule or in the surface of the solid. Hence the average concentration of force centers in the monolayer is less than that for the solid, and so the energy of the nonspecific dispersion interaction tends to be small. Here the support adsorbent must differ from the supports of gas—liquid chromatography in that it must be a strong (but nonspecific) adsorbent with adequate geometrical homogeneity and suitably high specific surface; all of these are needed to give the required column performance for the lower adsorption energy, and they are provided by graphitized carbon blacks of the Graphon type [28, 133] and by acetylene blacks [28, 134].

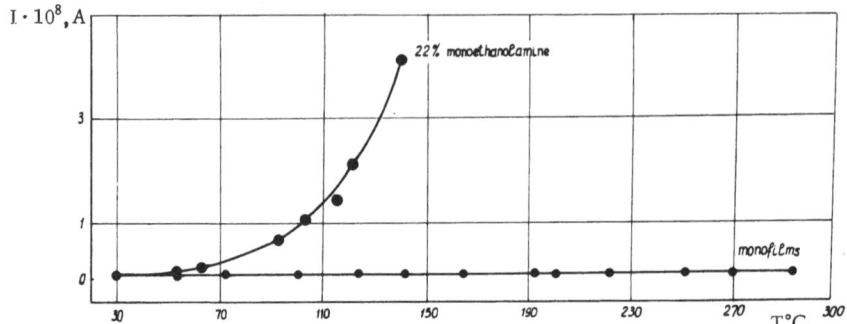

Fig. 33. Relation of background ion current to temperature for thick and monomolecular films of monoethanolamine on macroporous silica gel containing alumina.

This adsorption modification allows one to exploit one of the main advantages of gas–liquid chromatography, namely the chemical variety of fixed phases, while eliminating the disadvantages due to volatility and slow diffusion in the liquid films. Appropriate monolayers may be produced by adsorption of vapors or by adsorption of macromolecules from solution in volatile solvents. The latter method allows one to use compounds of high molecular weight. The monolayer gives only a low energy of nonspecific interaction, while the regular array of functional groups and hydrocarbon radicals provides high selectivity. A monolayer of molecules of ordinary size is strongly bound by the carrier adsorbent at high T if the molecules are bound chemically to the surface or the molecular adsorption is very strong. Figure 33 shows the background current as a function of T for: 1) macroporous silica gel containing alumina and bearing 22% of monoethanolamine as fixed liquid phase; 2) silica gel coated only with a monolayer of this compound. The vapor pressure is much reduced [162] by the specific interactions of the monolayer (strong hydrogen bonds from OH, ester, and amine groups to surface OH, chemisorption at aprotonic centers in the alumina or alumina–silica, and ester bonds to the silica).

Type III adsorbents are readily produced by deposition of a group B compounds (containing, say, CN or oxygen bridges) on a strong (but nonspecific) adsorbent, as for Graphon [5, 16]. The result of deposition of polyethylene glycol is a specific adsorbent of type III, which bears $-CH_2CH_2O-$ groups in which the oxygen has nonbonding pairs, which provide peripherally localized electron density.

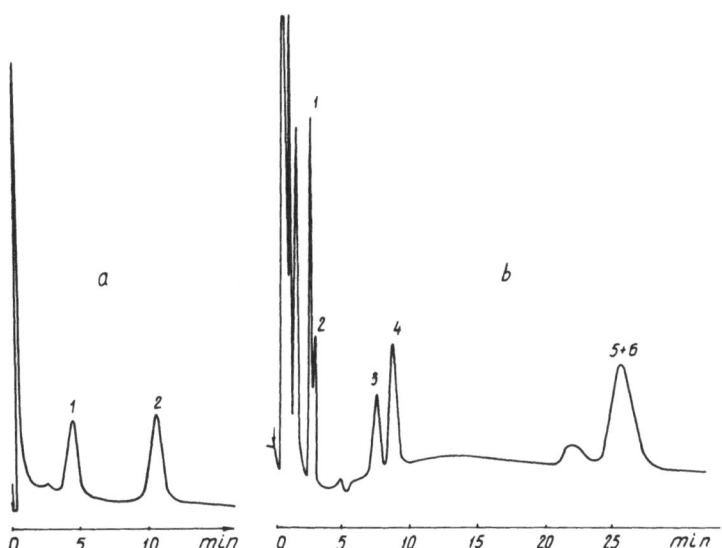

Fig. 34. Separation of (a) 1) etiocholanolone and 2) androsterone; (b) 1) phenanthrene; 2) anthracene; 3) fluoroanthene; 4) pyrene; 5) chrysene; 6) naphthacene. Column 2 m long at 302°C containing graphitized carbon black with 5% phthalocyanin.

Figure 32b shows that such a monolayer strongly retains group D compounds containing OH, alcohols being more strongly held than cyanides and nitro derivatives, although the latter have much larger dipole moments, because the oxygen groups in the glycol (Fig. 32b) have a dipole moment much less than that of the CN groups on chemically modified silica gel (Fig. 32a).

Type III thus provides great scope for adjustment of specificity and of the balance between specific and nonspecific interaction. In addition, type I or type II adsorbents may be produced by deposition of macromolecules of groups A or C and D, as for macroporous silica gel or graphitized carbon black. Deposition of solid aromatic compounds on graphitized carbon black gave [142, 143, 163-165] specific adsorbents of type III; in particular, treatment of Sterling MT-3100 graphitized black with anthraquinone gave a material that adsorbes nonspecifically not only hydrocarbons but also molecules with π-electrons and nonbonding pairs (aromatic hydrocarbons, ketones, ethers, alkyl chlorides). Mass transfer occurs very rapidly with such adsorbents. The lower average concentration of force centers means that the heats of adsorption are lower than for the parent black and are often below the heats of condensation, so separations can be performed with lower column temperatures.

Organic crystals and monolayers on graphitized carbon often show a specificity bearing a complicated relation to the structure. For instance, metal phthalocyanins [143] produce extensive conjugation and thus show relatively less specificity with respect to OH and NH groups, whereas they show high specificity for group B compounds, evidently on account of interaction of the latter with the metal.

Organic crystals on graphitized carbon black often show relatively low thermal stability; e.g., anthraquinone on carbon black is not stable above 160°C [164], so it is best to use monolayers of the most stable solid organic compounds; in particular phthalocyanins and polynuclear aromatics [143, 165], which are nonvolatile up to 300°C or more. Variation of the metal in the phthalocyanin provides control of the specificity and has given especially good results [143]. Figure 33 shows good separation of steroids on such an adsorbent, which is not attainable by gas–liquid chromatography; the same applies [143] to the separation of anthracene and phenanthrene.

A review [135] deals with the effects of the amount of liquid deposited on the macroporous base and of the consequent reduction in surface area.

Porous polymers are important new adsorbents for gas chromatography; several methods of production have been described [166-168]. Solvent sublimation [168] provides aerogels of amorphous polymers with s up to 200 m^2/g; polymerization of styrene with divinylbenzene as cross-linking agent in an inert medium [169-171] may also be used. The small pores with a high nonspecific interaction potential provide fairly high heats of adsorption and V_s for n-alkanes (close to those for graphitized carbon black).

COMPARISON OF ADSORBENTS OF TYPES I, II, and III

Nonspecific Adsorption of n-Alkanes. The n-alkanes are used in gas–liquid chromatography as standards for various parameters [136, 137] used in identification. They are particularly convenient as standards in gas-adsorption chromato-

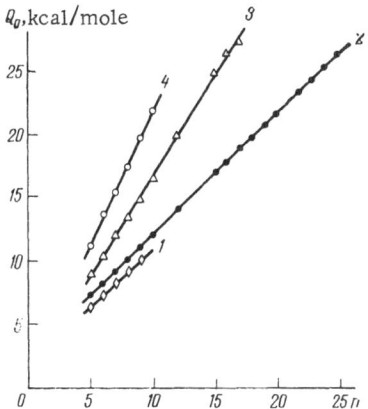

Fig. 35. Relation of gas-chromatographic Q_0
for n-alkanes to number of carbon atoms n
for adsorption on: 1) polyethylene glycol
monolayer; 2) macroporous silica gel; 3)
graphitized thermal carbon black; 4) crystals
of NaA zeolite.

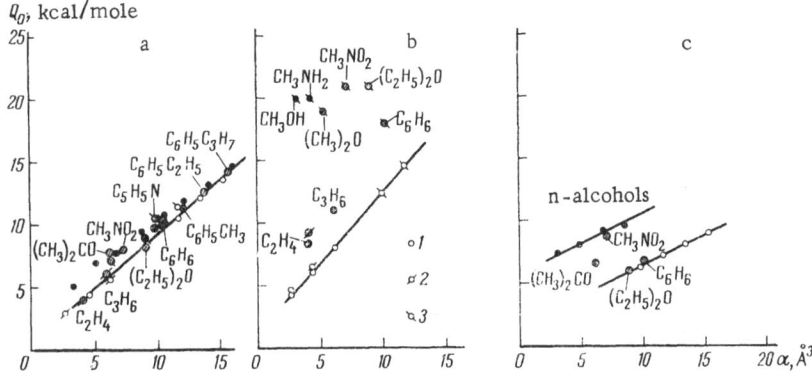

Fig. 36. Relation of Q_0 to polarizability α for: a) graphitized carbon black; b) channels
of NaX zeolite; c) monolayer of polyethylene glycol on Graphon for compounds of: O) group
A; ◑) group B; ●) group D; 1) gas-chromatographic data; 2) isosteric; 3) calorimetric.

graphy because their long molecules lie along the surface. The
classification (Table 1) puts them in group A (nonspecific only), so
they are convenient for comparing adsorbents as regards nonspecific adsorption.

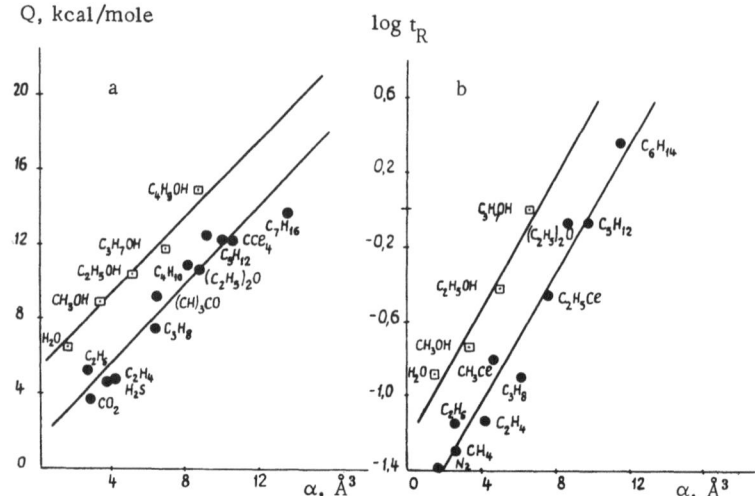

Fig. 37. Dependence on electronic polarizability α for chromatographic characteristics deduced with small samples for the porous polymer Chromosorb-102:
a) Q; b) $\log t_R$ [169].

Figure 35 shows the relation of the gas-chromatographic Q_0
for n-alkanes to number of carbon atoms n for adsorption on various adsorbents; Q_0 increases from Graphon coated with polyethylene glycol (nonspecific, weakly adsorbing, type III) to macroporous
silica gel (nonspecific, weakly adsorbing, type II), to graphitized
thermal carbon black (nonspecific, strongly adsorbing, type I), and
finally to the outer surface of NaA zeolite (nonspecific, strongly
adsorbing, type II) [128]. Q_0 increases linearly with n in all cases,
while the slopes (which represent the increments in Q_0 per CH_2
group) increase with the energy of the nonspecific interaction, i.e.,
with the surface density of carbon atoms (transition from polyethylene glycol monolayer to graphitized carbon black) or of oxygen
atoms (transition from silica gel to zeolite).

The increment $Q_{0.\ CH_2}$ itself increases by about 7% on going
from external adsorption on NaA zeolite to internal adsorption on
NaX zeolite [128].

The selectivity in nonspecific interaction may thus be adjusted
by suitable choice of surface.

Adsorption of Groups A, B, and D.   Specificity
is best examined by comparing V_s or Q_0 for n-alkanes (group A)

Table 6. Heats of Adsorption (kcal/mole) for Pairs
of Group A and B Molecules on Nonspecific and
Specific Adsorbents of Types I and II [5]

Compound	Adsorbent			
	Nonspecific		Specific	
	graphitized carbon black $(\theta = 0)$	dehydroxylated silica gel $(\theta = 0.5)$	hydroxylated silica gel $(\theta = 0.5)$	NaX $(\theta = 0)$
Argon	2.4	2.1	2.1	3.1
Nitrogen.	2.3	2.2	2.8	5.2
Ethane	4.3	4.2	4.4	6.2
Ethylene	4.0	3.8	5.2	9.2
n-Hexane	11.4	8.8	8.8	14.7
Benzene	10.1	8.6	10.2	18.0
n-Pentane	9.2	—	7.3	12.4
Diethyl ether . .	8.9	8.5	15.0	21.0

with those of the corresponding derivatives bearing functional groups
(groups B and D). Following Barrer [9, 138], we may plot Q_0 as a
function of the total polarizability α, which largely determines the
contribution from nonspecific interaction. Figure 36 [139, 140]
shows Q_0 as a function of α for n-alkanes, n-alkylbenzenes, n-nitroalkanes, n-alkyl cyanides, n-alcohols, and n-amines on graphitized
thermal carbon black (nonspecific, type I), on channels in NaX zeolite (specific, type II), and polyethylene glycol on Graphon (specific,
type III). Table 6 compares the Q_0 for various adsorbents of types
I and II for compounds of groups A and B similar in molecular size
and in α or in total number of carbon and oxygen atoms.

The specificity of Chromosorb-102 and Porapak P and Q may
be judged from Q as a function of electronic polarizability α (Fig.
37), these Q being derived from the narrow symmetrical peaks for
small samples. The relationship is nearly linear, with the points
for group B polar compounds almost on the same line as for group
A (n-alkanes, nonspecific interaction), although the former are capable of specific interaction with the positive charges localized at
the periphery. On the other hand, the Q for group D compounds,
which contain functional groups of OH type, are much larger for
the same value of α (i.e., for approximately the same energy of
nonspecific interaction). This is due to the benzene rings at the

surface of the porous polymer, whose π-electrons thus make a weakly specific adsorbent of type III, which adsorbs group D compounds more strongly.

Figure 37 shows $\log t_R$ as a function of α [169-171] (the Q are not given in the original papers, but $\log t_R$ is a linear function of Q for small samples and symmetrical peaks). This also shows that a single linear relation of $\log t_R$ to α applies for groups A and B. There is also a linear relation for group D, but the $\log t_R$ (and hence Q) are much larger. This shows that porous polymers (in particular, Chromosorb 102 and Porapak P and Q) may be assigned to type III (weakly specific); V_S is not very dependent on the boiling point or dipole moment, being governed mainly by α.

Porous polymers may be made with specifically adsorbing surfaces by copolymerization with compounds having appropriate polar functional groups, examples being Porapak Q, R, S, and T [174], and also polyacrylonitriles [183]. The t_R for group B compounds (diethyl ether, acetoacetic ester, acetone, nitromethane, acetonitrile) increase with the dipole moment for a polyacrylonitrile column, on account of the very high dipole moment of the CN group; α and the molecular weight have less effect. A strong dipole—dipole interaction with the CN group occurs for nitromethane and acetonitrile, which have the highest dipole moments in the above list. Group D compounds are adsorbed especially strongly by polyacrylonitrile; for instance, V_S and Q for n-butanol exceed those for diethyl ether and n-pentane, because the alcohol molecule has a specific (hydrogen-bond) interaction with the CN group in addition to the nonspecific dispersion interaction and the usual dipole—dipole interaction. Polyacrylonitrile is therefore a strongly specific type III adsorbent.

Figure 36 and Table 6 show that the capacity for specific interaction in groups B and D does not appear in adsorption on graphitized black and dehydroxylated silica gel, but it appears very clearly for specific type II adsorbents (hydroxylated silica gel and zeolite). Dehydroxylation of silica gel very largely suppresses the specific adsorption in group B; the material is thus very similar to nonspecific type I adsorbents.*

---

*Some group B and D compounds, e.g., alcohols, react chemically with dehydroxylated silica gel [80, 141].

The specific effects for group B appear in generally the same way for polar compounds (e.g., ether) and for nonpolar ones (benzene, ethylene, nitrogen), and this for hydroxylated silica gel and for cationized zeolite. Hence the presence of a dipole moment is not sufficient to produce specific interaction, which rather requires an appropriate peripheral electron-density distribution in adsorbate and adsorbent. This is confirmed by the fact that Q_0 for nitrobenzene on type II specific adsorbents is less than that for aniline, while Q_0 for nitromethane is less than that for methylamine, although the first member in each case has a dipole moment several times that of the second. This resembles the relation of adsorption to change of spectrum for specifically adsorbed compounds on hydroxylated surfaces of acid type and on cationized zeolites, which indicates that the hydrogen bond is only a particular case of specific interaction [2-5].

It is inadequate to classify compounds as polar and nonpolar in relation to specific interactions, which are not restricted to hydrogen bonds, either, though they remain molecular (the molecules retain their chemical individuality). This distinguishes specific molecular interactions from chemical interactions of donor–acceptor type, which involve complete charge transfer and loss of chemical individuality. The subject requires a more detailed quantum-mechanical consideration of peripheral electron densities and of the energies of specific molecular interactions.

The evidence presented in this chapter shows that suitable modification of the surface of a solid can provide a wide variety of nonspecific and specific adsorbents for gas chromatography; these should provide the required selectivity in conjunction with high performance and adequate thermal and chemical stabilities.

LITERATURE CITED

1.   A. V. Kiselev, Rev. Gen. Caoutchouc, 41:377 (1964).
2.   A. V. Kiselev, Zh. Fiz. Khim., 38:2753 (1964).
3.   A. V. Kiselev, in collection: Gas Chromatography, Transactions of the Third All-Union Conference on Gas Chromatography, Dzerzhinsk, Izd. Dzerzhinsk. Fil. OKBA, 1966, p. 15.
4.   A. V. Kiselev, in collection: Gas Chromatography 1964, A. Goldup (ed.), London, 1965, p. 238.
5.   A. V. Kiselev, Discussions Faraday Soc., 40:205 (1965).
6.   V. A. Barachevskii, V. E. Kholmogorov, and A. N. Terenin, Dokl. Akad. Nauk SSSR, 152:1143 (1963).

7. A. V. Kiselev, V. I. Lygin, and M. Sh. Rozenberg, Kinetika i Kataliz, 7:907 (1966).

8. A. V. Kiselev and A. V. Uvarov, Surface Sci., 6:401 (1967).

9. R. M. Barrer, J. Colloid. Interface Sci., 21:415 (1966).

10. D. H. Everett, in collection: Gas Chromatography 1964, A. Goldup (ed.), London, 1965, p. 219.

11. M. R. Basila, J. Chem. Phys., 35:1151 (1961).

12. R. M. Barrer, in collection: The Structure and Properties of Porous Materials, D. H. Everett and F. Stone (eds.), London, 1958, p. 6.

13. R. M. Barrer, in collection: Nonstoichiometric Compounds, L. Mondelcorn (ed.), New York, Academic Press, 1963, p. 309.

14. A. V. Kiselev and A. A. Lopatkin, Kinetika i Kataliz, 4:786 (1963).

15. C. G. Scott, in collection: Gas Chromatography 1962, M. van Swaay (ed.), London, 1962, p. 46.

16. L. D. Belyakova, A. V. Kiselev, N. V. Kovaleva, L. N. Rozanova, and V. V. Khopina, Zh. Fiz. Khim., 42:177 (1968).

17. N. M. Popov, V. I. Kasatochkin, and V. M. Luk'yanovich, Dokl. Akad. Nauk SSSR, 131:609 (1960).

18. D. Graham and W. S. Kay, J. Colloid Sci., 16:182 (1961).

19. A. V. Kiselev, Kolloidn. Zh., 20:388 (1958).

20. R. H. Fowler and E. A. Guggenheim, Statistical Thermodynamics [Russian translation], Moscow, IL, 1949 [English edition: Cambridge University Press].

21. T. L. Hill, J. Chem. Phys., 14:441 (1946).

22. J. H. de Boer, The Dynamical Character of Adsorption [Russian translation], Moscow, IL, 1962 [English edition: Oxford University Press, 1953].

23. R. A. Beebe, A. V. Kiselev, N. V. Kovaleva, R. F. S. Tyson, and J. M. Holmes, Zh. Fiz. Khim., 38:708 (1964).

24. A. G. Bezus, V. P. Dreving, and A. V. Kiselev, Zh. Fiz. Khim., 41:2937 (1967); A. G. Bezus, Dissertation, Mosk. Gos. Univ., 1966.

25. B. G. Aristov, V. Bosacek, and A.V. Kiselev, Trans. Faraday Soc., 63:2057 (1967).

26. R. M. Barrer and L. V. C. Rees, Trans. Faraday Soc., 57:999 (1961).

27. A. V. Kiselev, Yu. S. Nikitin, R. S. Petrova, K. D. Shcherbakova, and Ya. I. Yashin, Anal. Chem., 36:1526 (1964).

28. A. V. Kiselev, E. A. Paskonova, R. S. Petrova, and K. D. Shcherbakova, Zh. Fiz. Khim., 38:161 (1964).

29. A. V. Kiselev and Ya. I. Yashin, Zh. Fiz. Khim., 40:603 (1966).

30. A. V. Kiselev and Ya. I. Yashin, Zh. Fiz. Khim., 40:429 (1966).

31. N. N. Avgul', A. V. Kiselev, and I. A. Lygina, Izv. Akad. Nauk SSSR, Otd. Khim. Nauk, 1962, p. 32.

32. L. D. Belyakova, A. V. Kiselev, and N. V. Kovaleva, Zh. Fiz. Khim., 40:1494 (1966); A. V. Kiselev, I. A. Migunova, and Ya. I. Yashin, Zh. Fiz. Khim., 41:1235 (1967).

33. N. N. Avgul', A. V. Kiselev, and I. A. Lygina, Izv. Akad. Nauk SSSR, Otd. Khim. Nauk, 1961, p. 1395.

34. A. V. Kiselev and Ya. I. Yashin, in collection: Gas Chromatographie 1965, H. P. Angele and H. G. Struppe (eds.), Berlin, Akademie Verlag, 1965, p. 43.

35. A. V. Kiselev, I. A. Migunova, and Ya. I. Yashin, Neftekhimiya, 7:807(1967).

36. H. W. Dauben and K. S. Pitzer, in collection: Steric Effects in Organic
     Chemistry [Russian translation], Moscow, IL, 1960, p. 28 [English edition:
     M. S. Newman (ed.), New York, Wiley].

37. A. V. Kiselev, G. M. Petov, and K. D. Shcherbakova, Zh. Fiz. Khim., 41:
     1418 (1967); G. M. Petov and K. D. Shcherbakova, in collection: Gas Chro-
     matography 1966, A. B. Littlewood (ed.), London, Institute of Petroleum,
     1967, p. 50.

38. A. V. Polyakova, G. M. Sal'nikova, and Ya. I. Yashin, in collection: Gas
     Chromatography, No. 5, Moscow, Izd. NIITÉKhim, 1967.

39. R. M. Barrer, Proc. Roy. Soc., A161:476 (1937).

40. S. Brunauer, The Adsorption of Gases and Vapors, Vol. 1, Princeton University
     Press, 1945.

41. B. V. Il'in, Nature of Adsorption Forces, Moscow, Gostekhteorizdat, 1952.

42. A. D. Crowell, J. Chem. Phys., 22:1397 (1954).

43. A. D. Crowell, J. Chem. Phys., 26:1407 (1957).

44. A. D. Crowell, J. Chem. Phys., 29:446 (1958).

45. A. D. Crowell and W. A. Steele, J. Chem. Phys., 34:1347 (1961).

46. D. M. Young and A. D. Crowell, Physical Adsorption of Gases, London, 1962.

47. N. N. Avgul', A. A. Isirikyan, A. V. Kiselev, I. A. Lygina, and D. P. Poshkus,
     Izv. Akad. Nauk SSSR, Otd. Khim. Nauk, 1957, p. 1314.

48. N. N. Avgul', A. V. Kiselev, I. A. Lygina, and D. P. Poshkus, Izv. Akad. Nauk
     SSSR, Otd. Khim. Nauk, 1959, p. 1196.

49. N. N. Avgul', A. V. Kiselev, and I. A. Lygina, Izv. Akad. Nauk SSSR, Otd.
     Khim. Nauk, 1962, p. 32.

50. N. N. Avgul', A. V. Kiselev, and I. A. Lygina, Izv. Akad. Nauk SSSR, Otd.
     Khim. Nauk, 1961, p. 1404.

51. A. V. Kiselev and D. P. Poshkus, Trans. Faraday Soc., 59:176 (1963).

52. A. V. Kiselev and D. P. Poshkus, Trans. Faraday Soc., 59:428 (1963).

53. N. N. Avgul', A. V. Kiselev, and I. A. Lygina, Trans. Faraday Soc., 59:2113
     (1963).

54. J. R. Sams, G. Constabaris, and G. D. Halsey, J. Phys. Chem., 64:1689 (1960).

55. G. Constabaris, J. R. Sams, and G. D. Halsey, J. Phys. Chem., 65:367 (1961).

56. J. R. Sams, Trans. Faraday Soc., 60:149 (1964).

57. A. V. Kiselev, D. P. Poshkus, and A. Ya. Afreimovich, Zh. Fiz. Khim., 38:
     1514 (1964).

58. A. V. Kiselev, D. P. Poshkus, and A. Ya. Afreimovich, Zh. Fiz. Khim., 39:
     1190 (1965).

59. D. P. Poshkus, Zh. Fiz. Khim., 39:2962 (1965); A. V. Kiselev, in collection:
     Gas Chromatography, Transactions of the Third All-Union Conference on Gas
     Chromatography, Dzerzhinsk, Izd. Dzerzhinsk. Fil. OKBA, 1966, p. 145.

60. D. P. Poshkus, Discussions Faraday Soc., 40:195 (1965).

61. A. V. Kiselev and D. P. Poshkus, Zh. Fiz. Khim., 41:2647 (1967).

62. G. Kirkwood, Phys. Z., 33:57 (1932).

63. A. Müller, Proc. Roy. Soc., A154:624 (1936).

64. A. V. Kiselev and D. P. Poshkus, Zh. Fiz. Khim., 32:2824 (1958).

65. L. A. Girifalco and R. A. Lad, J. Chem. Phys., 25:693 (1956).

66.   D. P. Poshkus and A. Ya. Afreimovich, Abstracts of Papers at the Fourth All-Union Conference on Gas Chromatography, Moscow, 1966, p. 13; Zh. Fiz. Khim., Vol. 42 (1968).

67.   L. D. Belyakova and A. V. Kiselev, Izv. Akad. Nauk SSSR, Otd. Khim. Nauk, 1966, p. 638.

68.   A. V. Kiselev, A. Ya. Korolev, R. S. Petrova, and K. D. Shcherbakova, Kolloidn. Zh., 22:671 (1960).

69.   A. V. Kiselev, Vestn. Mosk. Gos. Univ., Ser. Khim., 1961, p. 29.

70.   I. Yu. Babkin and A. V. Kiselev, Zh. Fiz. Khim., 36:2448 (1962).

71.   A. V. Kiselev, R. S. Petrova, and K. D. Shcherbakova, Kinetika i Kataliz, 5:526 (1964).

72.   I. V. Borisenko, A. V. Kiselev, R. S. Petrova, V. K. Chuikina, and K. D. Shcherbakova, Zh. Fiz. Khim., 39:2685 (1965).

73.   A. V. Kiselev, V. K. Chuikina, and K. D. Shcherbakova, Zh. Fiz. Khim., 40:140 (1966).

74.   J. Bohemen, S. H. Langer, R. H. Perrett, and J. H. Purnell, J. Chem. Soc., (1960), p. 2444.

75.   R. H. Perrett and J. H. Purnell, J. Chromatog., 7:455 (1962).

76.   G. A. Galkin, S. P. Zhdanov, A. V. Kiselev, and I. A. Lygina, Kolloidn. Zh., 25:123 (1963).

77.   A. V. Kiselev, in collection: Gas Chromatography, Transactions of the First All-Union Conference on Gas Chromatography, Moscow, Izd. Akad. Nauk SSSR, 1960, p. 45.

78.   I. Yu. Babkin and A. V. Kiselev, Dokl. Akad. Nauk SSSR, 129:357 (1959).

79.   R. K. Iler, Colloid Chemistry of Silica and Silicates, Ithaca, New York, Cornell University Press, 1955.

80.   L. D. Belyakova and A. V. Kiselev, Zh. Fiz. Khim., 33:1534 (1959).

81.   C. C. Ballard, E. C. Broge, R. K. Iler, D. S. St. John, and J. R. McWhorter, J. Phys. Chem., 65:20 (1961).

82.   B. G. Aristov, I. Yu. Babkin, and A. V. Kiselev, Kolloidn. Zh., 24:643 (1962).

83.   I. V. Borisenko, N. I. Bryzgalova, T. B. Gavrilova, and A. V. Kiselev, Nefte-khimiya, 6:129 (1966).

84.   A. V. Kiselev, Dokl. Akad. Nauk SSSR, 106:1046 (1956).

85.   R. Haldeman and P. H. Emmett, J. Am. Chem. Soc., 78:2917 (1956).

86.   J. J. Fripiat and J. Uytterhoeven, J. Phys. Chem., 66:800 (1962).

87.   V. Ya. Davydov, A. V. Kiselev, and L. T. Zhuravlev, Trans. Faraday Soc., 60:2254 (1964).

88.   L. T. Zhuravlev and A. V. Kiselev, Zh. Fiz. Khim., 39:453 (1965).

89.   F. H. Hambleton, J. A. Hockey, and J. A. G. Taylor, Trans. Faraday Soc., 62:801 (1966).

90.   J. Uytterhoeven, M. Slex, and J. J. Fripiat, Bull. Soc. Chim. France, No. 6: 1800 (1965).

91.   V. M. Chertov, D. B. Dzhambaeva, A. S. Plachinda, and I. E. Neimark, Zh. Fiz. Khim., 40:520 (1966).

92.   G. K. Boreskov, M. S. Borisova, O. M. Dzhigit, V. A. Dzis'ko, V. P. Dreving, A. V. Kiselev, and O. A. Likhacheva, Zh. Fiz. Khim., 22:603 (1948).

93. G. K. Boreskov, M. S. Borisova, V. A. Dzis'ko, A. V. Kiselev, O. A. Lik-
     hacheva, and T. N. Morokhovets, Dokl. Akad. Nauk SSSR, 62:649 (1948).
94. N. V. Akshinskaya, A. V. Kiselev, and Yu. S. Nikitin, Zh. Fiz. Khim., 37:
     927 (1963).
95. A. V. Kiselev and V. I. Lygin, Usp. Khim., 31:351 (1962).
96. L. Little, Infrared Spectra of Adsorbed Species, London, Academic Press, 1966.
97. R. J. McDonald, J. Am. Chem. Soc., 79:850 (1957).
98. R. J. McDonald, J. Phys. Chem., 62:1168 (1958).
99. V. Ya. Davydov and A. V. Kiselev, Zh. Fiz. Khim., 37:2593 (1963).
100. V. Ya. Davydov, A. V. Kiselev, and B. V. Kuznetsov, Zh. Fiz. Khim., 39:
     2058 (1965).
101. A. N. Terenin, in collection: Role of Surface Compounds in Adsorption,
     A. V. Kiselev (ed.), Moscow, Izd. MGU, 1957, p. 206.
102. V. Ya. Davydov, A. V. Kiselev, and V. I. Lygin, Dokl. Akad. Nauk SSSR,
     147:131 (1962).
103. A. V. Kiselev, Surface Sci., 3:292 (1965).
104. W. G. Gordy, J. Chem. Phys., 7:93 (1939).
105. A. V. Iogansen, Dokl. Akad. Nauk SSSR, 164:610 (1965).
106. G. A. Galkin, A. V. Kiselev, and V. I. Lygin, Zh. Fiz. Khim., 40:2880 (1966).
107. L. D. Belyakova and A. V. Kiselev, Dokl. Akad. Nauk SSSR, 119:298 (1958).
108. G. A. Galkin, S. P. Zhdanov, A. V. Kiselev, and I. A. Lygina, Zh. Fiz. Khim.,
     37:228 (1963).
109. B. G. Aristov and A. V. Kiselev, Zh. Fiz. Khim., 38:1984 (1964).
110. A. G. Bezus and A. V. Kiselev, Zh. Fiz. Khim., 40:580 (1966).
111. A. V. Kiselev, V. K. Chuikina, and K. D. Shcherbakova, Zh. Fiz. Khim., 40:
     1533 (1966).
112. O. M. Dzhigit, A. V. Kiselev, and G. G. Muttik, Kolloidn. Zh., 23:504 (1961).
113. A. V. Kiselev and B. A. Frolov, Kinetika i Kataliz, 3:767 (1962).
114. A. V. Kiselev, Yu. L. Chernen'kova, and Ya. I. Yashin, Neftekhimiya, 5:589
     (1965).
115. V. Bosaček, in collection: Synthesis, Properties, and Uses of Zeolites, Mos-
     cow, Izd. Nauka, 1965, p. 103.
116. O. M. Dzhigit, K. Karpinskii, A. V. Kiselev, T. A. Mel'nikova, K. N. Mikos,
     and G. G. Muttik, Zh. Fiz. Khim., 42:198 (1968).
117. O. M. Dzhigit, S. P. Zhdanov, A. V. Kiselev, V. K. Chuikina, K. N. Mikos,
     and G. G. Muttik, Zh. Fiz. Khim., 41:1431 (1967).
118. O. M. Dzhigit, S. P. Zdhanov, and K. N. Mikos, in collection: Synthesis,
     Properties, and Uses of Zeolites, Moscow, Izd. Nauka, 1965, p. 46.
119. R. M. Barrer and G. C. Bratt, Phys. Chem. Solids, 12:154 (1959).
120. A. V. Kiselev and Ya. I. Yashin, Zh. Fiz. Khim., 40:944 (1966).
121. A. G. Bezus, V. P. Dreving, and A. V. Kiselev, Zh. Fiz. Khim., 38:2924
     (1964).
122. S. Ross, J. K. Saelens, and J. P. Olivier, J. Phys. Chem., 66:696 (1962).
123. R. M. Barrer and J. W. Sutherland, Proc. Roy. Soc., A237:439 (1956).
124. P. E. Eberly, J. Phys. Chem., 66:812 (1962).
125. O. M. Dzhigit, A. V. Kiselev, and G. G. Muttik, Kolloidn. Zh., 25:34 (1963).

126. N. N. Avgul', A. V. Kiselev, A. A. Lopatkin, I. A. Lygina, and M. V. Serdobov, Kolloidn. Zh., 25:129 (1963).
127. H. W. Habgood, Can. J. Chem., 42:2340 (1964).
128. V. L. Keibal, A. V. Kiselev, I. M. Savinov, V. L. Khudyakov, K. D. Shcherbakova, and Ya. I. Yashin, Zh. Fiz. Khim., 41:2234 (1967).
129. A. G. Altenau and L. V. Rogers, Anal. Chem., 37:1432 (1965).
130. C. G. Scott and C. S. G. Phillips, Gas Chromatography 1964, A. Goldup (ed.), London, 1965, p. 2.
131. J. A. Favre and L. R. Kallenbach, Anal. Chem., 36:63 (1964).
132. A. V. Kiselev and K. D. Shcherbakova, in collection: Gas Chromatographie 1961, Berlin, Akademie Verlag, 1962, pp. 207 and 241.
133. W. D. Schaeffer, W. R. Smith, and M.H. Polley, J. Phys. Chem., 57:469 (1953).
134. A. V. Kiselev, N. V. Kovaleva, and R. S. Petrova, Kolloidn. Zh., 27:822 (1965).
135. S. Ross and E. D. Tolls, Summaries of Papers at the Twentieth International Congress on Theoretical and Applied Chemistry, Moscow, Izd. Nauka, 1965, A49, p. 50.
136. E. Kovats, Z. Anal. Chem., 181:351 (1961).
137. L. S. Ettre, Open Tubular Columns in Gas Chromatography, New York, Plenum Press, 1965.
138. R. M. Barrer, Discussions Faraday Soc., 40:231 (1965).
139. A. V. Kiselev, Discussions Faraday Soc., 40:229 (1965).
140. N. N. Avgul', A. V. Kiselev, L. Ya. Kurdyukova, and M. V. Serdobov, Zh. Fiz. Khim., 42:188 (1968).
141. W. Stober, Kolloid.-Z., 145:17 (1956).
142. C. Vidal-Madijar and G. Guiochon, Compt. Rend. Acad. Sci., Paris, 265:26 (1967).
143. C. Vidal-Madijar and G. Guiochon, in collection: Proceedings of the Sixth Symposium of Gas Chromatography, Berlin, 1968; J. Chromatog. (in press).
144. F. J. Wilkins, Proc. Roy. Soc., A164:496 (1938).
145. A. V. Kiselev, Zh. Fiz. Khim., 41:2470 (1967).
146. R. M. Barrer and B. Coughlan, Molecular Sieves, London, 1967.
147. N. N. Avgul', A. S. Guzenberg, A. V. Kiselev, L.Ya. Kurdyukova, and A. M. Ryabkin, Zh. Fiz. Khim. (in press).
148. A. S. Boikova and K. D. Shcherbakova, Neftekhimiya, 7:451 (1967).
149. W. Schneider, W. Brudereck, and I. Halasz, Anal. Chem., 36:1533 (1964).
150. D. P. Poshkus, in collection: Proceedings of the Sixth Symposium on Gas Chromatography, Berlin, 1968; J. Chromatog. (in press).
151. A. V. Kiselev, D. P. Poshkus, and A. Ya. Afreimovich, Zh. Fiz. Khim. (in press).
152. A. V. Kiselev and D. P. Poshkus, Zh. Fiz. Khim., 43:285 (1969).
153. A. V. Kiselev, in collection: Proceedings of the Sixth Symposium on Gas Chromatography, Berlin, 1968; J. Chromatog. (in press).
154. G. Constabaris, J. R. Sams, and G. D. Halsey, J. Phys. Chem., 65:367 (1961).
155. G. C. Goretti, A. Liberti, and G. Nota, J. Chromatog., 34:96 (1968).

156.   A. V. Kiselev, K. I. Sakodynskii, V. L. Khudyakov, and Ya. I. Yashin, Gas Chromatography, Izd. NIITÉKhim (in press).
157.   N. N. Avgul' and A. V. Kiselev, in collection: Chemistry and Physics of Carbon, Vol. 6 P. L. Walker (ed.). New York, Marcel Dekker, Inc.
158.   N. K. Bebris, A. V. Kiselev, and Yu. S. Nikitin, Kolloidn. Zh., 29:326 (1967).
159.   N. K. Bebris, G. E. Zaitseva, A. V. Kiselev, Yu. S. Nikitin, and Ya. I. Yashin, Neftekhimiya, 8(3):481 (1968).
160.   Yu. S. Nikitin, in collection: Proceedings of the Sixth Symposium on Gas Chromatography, Berlin, 1968; J. Chromatog. (in press).
161.   R. Day, A. V. Kiselev, B. V. Kuznetsov, and Yu. S. Nikitin, Kinetika i Kataliz (in press).
162.   V. I. Kalmanovskii, A. V. Kiselev, G. G. Sheshenina, and Ya. I. Yashin, Gas Chromatography, Moscow, Izd. NIITÉKhim, Issue 6, 1967, p. 45.
163.   C. Vidal-Madijar and G. Guiochon, Bull. Soc. Chim. France, 1966, p. 1096.
164.   C. Vidal-Madijar and G. Guiochon, Separation Sci., 2:1551 (1967).
165.   C. Vidal-Madijar and G. Guiochon, Nature, 215:1372 (1967).
166.   J. C. Moore, J. Polymer Sci., A2:835 (1964).
167.   J. C. Moore and J. C. Hendrickson, J. Polymer Sci., C8:233 (1965).
168.   G. V. Vinogradov, L. V. Titkova, N. V. Akshinskaya, N. K. Bebris, A. V. Kiselev, and Yu. S. Nikitin, Zh. Fiz. Khim., 40:84 (1966).
169.   O. L. Hollis, Anal. Chem., 38:309 (1966).
170.   O. L. Hollis and W. V. Hayes, J. Gas Chromatog., 4:235 (1966).
171.   O. L. Hollis and W. V. Hayes, in collection: Gas Chromatography 1965, A. B. Littlewood (ed.), Institute of Petroleum, 1966.
172.   A. Zlatkins and H. R. Kaufman, J. Gas Chromatog., 4:240 (1966).
173.   G. Bogart, Bull. Facts and Methods, 5:7 (1966).
174.   A. Klein, Varian Aerograph Technical Bulletin, No. 128 (1966).
175.   K. Derge, Fette, Seifen, Anstrichmittel, 69:407 (1967).
176.   S. Spenser, Bull. Facts and Methods, 8:2 (1967).
177.   K. Jones, J. Gas Chromatog., 8:432 (1967).
178.   J. Kikuchi, T. Kikkawa, and B. Kato, J. Gas Chromatog., 5:261 (1967).
179.   T. N. Gvozdovich, M. P. Kovaleva, G. K. Petrova, and Ya. I. Yashin, Neftekhimiya, 8:124 (1968).
180.   K. I. Sakodynskii and L. I. Moseva, Gas Chromatography, Moscow, Izd. NIITÉKhim, Issue 7, 1968, p. 18.
181.   L. I. Moseva, M. P. Kovaleva, and K. I. Sakodynskii, Gas Chromatography, Moscow, Izd. NIITÉKhim, Issue 7, 1968, p. 60.
182.   L. I. Moseva and K. I. Sakodynskii, Gas Chromatography, Moscow, Izd. NIITÉKhim, Issue 9, 1968.
183.   N. K. Bebris, A. V. Kiselev, and Yu. S. Nikitin, Neftekhimiya (in press).

Chapter III

Role of the Geometrical Structure of Adsorbents

THE GENERAL RELATION BETWEEN THE PERFORMANCE OF A COLUMN AND THE GEOMETRICAL STRUCTURE OF THE ADSORBENT

The choise of gross geometrical structure (specific surface and porosity) for a given surface composition is governed by the mixture to be separated. The lifetimes in the adsorbed state are small for gases and light hydrocarbons at ordinary temperatures, so the adsorbent must have a reasonably large specific surface. On the other hand, ordinary (or slightly elevated) temperatures are sufficient to prevent substantial peak broadening from surface inhomogeneity and pore exchange for gases (including light hydrocarbons) on amorphous adsorbents of high specific surface. Zeolites,* microporous silica gels, microporous glasses, and capillary glass columns with porous surfaces are used. For example, hydrogen isotopes and isomers have been separated on zeolites in a capillary column [1] and on glass capillary columns with a porous layer on the walls [2].

It is difficult to use strong specific adsorbents (zeolites) to separate compounds showing strong specific adsorption, such as ammonia, the lower amines, cyanides, and esters. Good results are provided here by nonspecific adsorbents with strong dispersion forces and uniform surfaces, such as Graphon channel black and

---

*See chapter V for a survey of the use of zeolites in analytical gas chromatography.

acetylene black. These nonspecific adsorbents may also be used to separate light hydrocarbons. Nonspecific adsorbents with particular uniform surfaces may also be used to separate water ahead of other components. The usual silica gels may be used to separate compounds that are largely nonspecifically adsorbed. The specific surface must be reduced and the pore size increased as the molecular size increases; compounds showing strong specific adsorption should be used with nonspecific adsorbents of highly uniform surface at elevated temperatures.

There are thus important effects from the gross structure, specific surface, surface inhomogeneity, and pore-size distribution. Here a classification is of value, as for the chemical composition of the surface.

CLASSIFICATION OF ADSORBENTS
BY GEOMETRICAL STRUCTURE

Kiselev in 1948 [3-5] gave such a classification, four basic types being distinguished in order of decreasing pore size and mode of distribution.

Type I, Nonporous. Examples are sodium chloride, graphitized carbon blacks, and nonporous amorphous adsorbents such as aerosil and thermal ungraphitized carbon blacks. The specific surfaces of these range up to hundreds of m^2/g. Such an adsorbent is mounted in the macropores of a support [6-8], aggregated into spheres (carbon blacks [9-11]), or pressed into grains of the required sizes (various crystalline materials). Between the particles of nonporous adsorbent there are then holes, i.e., a certain porosity; the sizes of these pores are usually comparable with those of the primary particles, and so exchange is rapid.

The uptake by unit surface is almost independent of the specific surface if the surface is uniform in composition [12-14], as is clear from V_s [see (II, 2) and Fig. 7] [7, 15].

The capillary condensation in these pores is usually slight for a freely poured nonporous adsorbent; it occurs only at relative vapor pressures close to one. Marked capillary-condensation hysteresis occurs after pressing of materials of s from 10 to 100 m^2/g [5, 6-18]; pressed discs of nonporous materials constitute a macroporous adsorbent (type II) whose pore shape and size are determined by the pressure as well as by the shape and size of the primary particles [17-19].

Type II, Homogeneous Macroporous. These include xerogels, macroporous glasses, and pressed nonporous powders of size over 100 Å and s less than 300-400 m^2/g. The structural similarity between the pores of xerogels and those of pressed nonporous spheroidal particles is due to the corpuscular (in particular, globular) structure of the skeleton in a xerogel [5, 12, 20-25]. These materials show pronounced capillary-condensation hysteresis. A xerogel may be made directly as microspheres of size suitable for column filling (~0.2 mm). Normal commercial macroporous silica gels usually have s of at least 300 m^2/g and pore sizes only 100-200 Å, and hence for gas chromatography of substances of medium or (especially) high boiling point it is important to reduce s and widen the pores. The production of macroporous specimens from ordinary silica gels is considered below.

Type III, Homogeneous Microporous. Examples are amorphous microporous xerogels [3, 4, 26-28], microporous glasses [29], many specimens of activated charcoal [30] (e.g., Saran type [31]), and porous crystals, e.g., A and X zeolites [32-35]. The openings of the pores in a porous crystal are identical, so these are particularly convenient for molecular-sieve separation [32, 33, 35-38]. If a compound cannot enter the pores, the crystal behaves as a type I (nonporous) adsorbent, as in the use of NaA zeolite as a nonporous adsorbent for hydrocarbons (p. 50).

Type IV, Inhomogeneously Porous. These are often xerogels, e.g., chalky silica gels produced by precipitation from a silicate solution by hydrolyzed salts of strong acids [20]. They usually contain many strongly adsorbing fine pores, and hence they do not find such extensive use in gas chromatography as do types I-III.

GEOMETRICAL MODIFICATION OF ADSORBENTS

Heating of an ordinary silica gel to 700-950°C (dependent on the pore size) produces sintering of the skeleton, so the pore volume is much reduced [27, 39-41]. Water vapor at 700-800°C [21, 23, 24, 40, 41], especially hydrothermal treatment in an autoclave [25, 42, 43], at first produces only a slight change in pore volume,

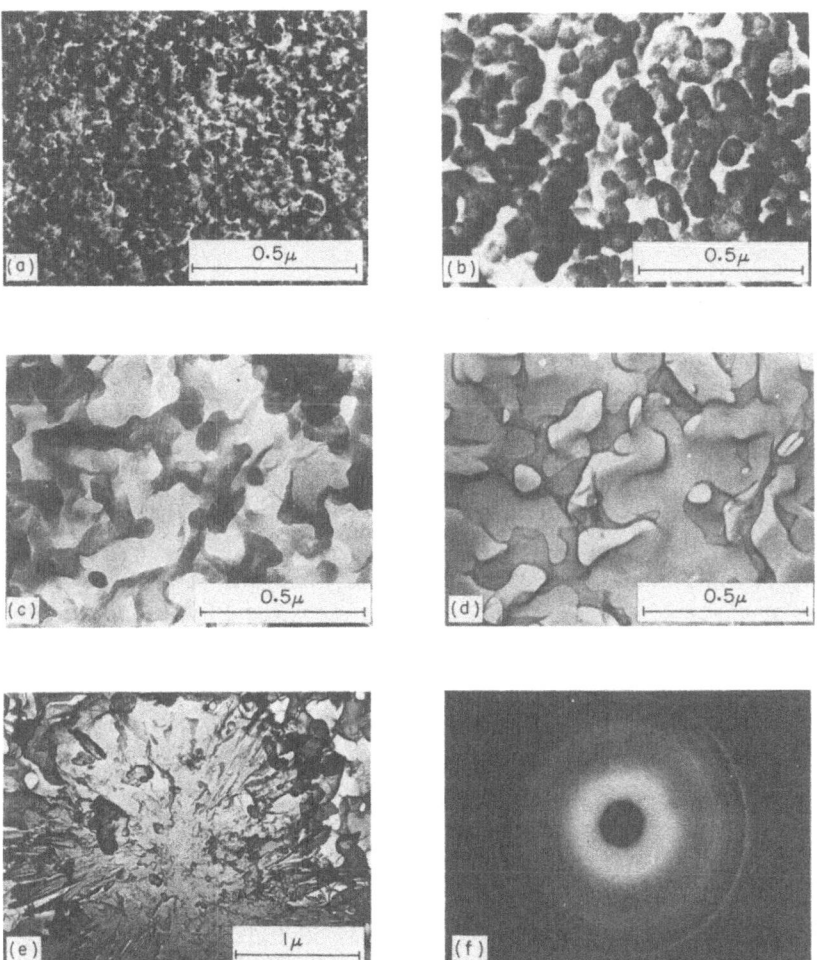

Fig. 38. Electron micrographs of shadowed carbon replicas: a) initial silica gel, s about 340 m^2/g, dominant particle size D about 100 Å, dominant pore size d about 100 Å; b) after hydrothermal treatment at 250°C (corpuscular structure still present, s about 35 m^2/g, D about 800 Å, d about 800 Å); c) after hydrothermal treatment at ~ 350°C (dentate structure, s ~5 m^2/g); d) after hydrothermal treatment at 370° (onset of crystallization); e) transmission picture of resulting crystals; f) electron-diffraction pattern.

while s is substantially reduced. This occurs by loss of small globules and substantial growth of the large ones [21, 23-25], and so the pores are enlarged [15, 25, 40-43]. In this way s is readily reduced to 50-25 m^2/g, while the pores are enlarged to thousands

of Å [25, 41, 43]. Electron micrographs of shadowed carbon replicas (Fig. 38a and b) reveal the enlargement of the large grains,
which consist of intergrown globules [25]. The structure becomes
dentate (Fig. 38c and d) if the temperature and pressure of the
steam are raised further. The porosity is then gradually lost, and
the gel turns into spherulites (Fig. 38e) with a characteristic crystalline structure (Fig. 38f).

Hydrothermally treated silica gel contains much water and
structural OH in the volume and on the surface. Firing removes
much of the water, which leaves behind ultrapores, which accom-
pany the macropores (gaps between large particles). These ultra
pores slowly take up water vapor, and also some other compounds
with small molecules (CH_3OH, N_2), the uptake being inversely related to the molecular size [44-46]. This bidisperse structure
makes the material inconvenient for use in gas-chromatography
columns. The ultrapores are removed by firing, with the temperature rising gradually to 900-1000°C [44], and also by steam treatment at about 750°C [46]. In the first case the surface loses most
of its OH. If necessary, the specific adsorption can be increased
by prolonged boiling in water or by hydrothermal treatment under
mild conditions [46, 47].

Commercial silica gels tend to be of low chemical purity and
often contain, in particular, Al_2O_3 and Fe_2O_3, which produce chemical activity and substantial surface inhomogeneity, with resulting
loss of symmetry in peaks for polar compounds, as well as possible catalytic activity. Pure materials (aerosilogels) [170] contain <0.1% impurity and give much better peak symmetry. Aerosilogel is a coarse-pored and chemically very pure silica with a
surface of high geometrical and chemical homogeneity; the uses in
gas chromatography have been described [171-173]. High-temperature dehydroxylation of impure silicas gives rise to centers that
chemisorb many compounds of groups B and D, but that hardly af-
fect the adsorption of group A compounds [174].

PRODUCTION OF POROUS POLYMERS

Porous mineral adsorbents retaining the gel structure are
produced [48-50] by vacuum sublimation of the frozen intermicelle
liquid; this eliminates the surface-tension forces that tend to make

TABLE 7. Geometrical Characteristics of Porapak Q, R, S, and T

Porapak	s, m²/g	v, cm³/g	d, A
Q	634	1.185	74.8
R	547	1.035	75.6
S	536	1.017	76.0
T	306	0.701	91.4

the gel contract on normal drying. The method may also be used to prepare organic and heteroorganic polymers with wide pores. It has been shown [49, 50] that polymer aerogels may be made in this way. A basic condition for producing a high specific surface, at least for an amorphous polymer, is a temperature low enough for the polymer to remain in the vitrified state, because the motion of the macromolecules is inhibited below the vitrification point. Then the rigidity of the polymer molecules after removal of the solid solvent produces a stable spatial skeleton. The resulting polymer aerogel reflects more or less closely the structure of the polymer in solution.

Aerogels have been made by distillation of frozen solvent from block polystyrene and polyphenyldisiloxane [50]. The solution is frozen and put under vacuum. When most of the solvent has been removed, the temperature is raised to normal over a period of several hours, with continuous pumping.

These organic aerogels are difficult to use in gas chromatography, on account of swelling in the presence of organic vapors. The best results have been obtained with organosilicon aerogel [50]. When various methods of cross-linking are used, the products are extremely homogeneous macroporous adsorbents with regularly disposed surface functional groups. Hollis [51] has used such a porous polymer in gas-adsorption chromatography. Hollis has shown that s for porous polymers based on styrene and divinylbenzene may range from 1 to 660 m²/g; he has used specimens with s of 100-600 m²/g. The geometrical structure has been examined [175] for Porapak Q, R, S, and T (Waters Associates Inc., Framingham, Mass.), together with measurement of s, mean pore diameter d, micropore volume v, and pore-size distribution (Table 7). The pore structure is rather inhomogeneous, and there are many very small pores.

It is clear that Porapak T has an s much smaller than those
for types Q, R, and S; this occurs because Porapak T has more
large pores. It has been concluded [175] that the separating power
of these porous polymers is largely independent of the geometrical
structure but is much more dependent on the chemical nature of
the surface.

STRUCTURE OF ADSORBENT PORES
AND RETENTION OF MATERIALS

The uptake of adsorption [4, 12, 30, 52, 53] is greatly dependent on the pore structure and on the nature of the surface. Static
measurements of adsorption isotherms for hydrocarbons on silica
gel [12, 52-55] show that the uptake and heat of adsorption are inversely related to pore size but directly related to the number of
carbon atoms in the molecule. Figure 39 shows that the isotherms
become strongly curved as the pores grow narrower and are roughly
the same for different specimens with pores of about the same size
[4, 53, 54].

Vyakhirev et al. [56] have examined the effects of geometrical
structure on separation of gaseous hydrocarbons for microporous
silica gel. The completeness of separation is directly related to s
and inversely to pore size. The effects of pore size have also been
examined for liquids [15, 57]; the V_m for normal hydrocarbons were
reckoned per unit surface area for gels of different pore sizes but
identical surface composition [15, 57-59].

V_s is determined by the standard change in chemical potential of the adsorbate [60, 61]:

$$\Delta\mu_0 = -Q_0 - T\Delta S_0,$$

(III, 1)

in which Q_0 is the initial heat of adsorption and ΔS_0 is the corresponding standard differential entropy change. The ΔS_0 for molecules similar in structure do not differ very greatly, so the V_s are
governed mainly by Q_0.

Gas-chromatographic measurement of Q_0 for hydrocarbons
on silica gels shows (Fig. 40) that Q_0 is inversely related to pore
size, on account of increase in the potential of the nonspecific interaction [12, 52-55]; but for d > 500 Å there is hardly any effect on
Q_0 or V_s, no matter what s may be.

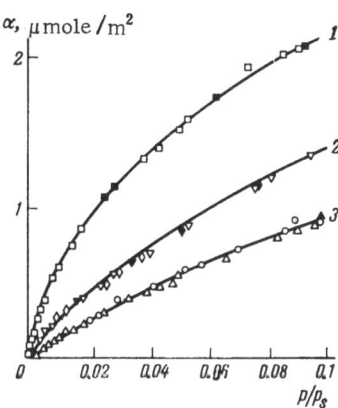

Fig. 39. Effects of pore size in silica gel on the surface concentration α for n-pentane and on the shape of the isotherm (p/p_s is the relative vapor pressure): 1) pore diameter 25 Å; 2) pore diameters of 36 and 38 Å; 3) pore diameter 100 Å and nonporous quartz.

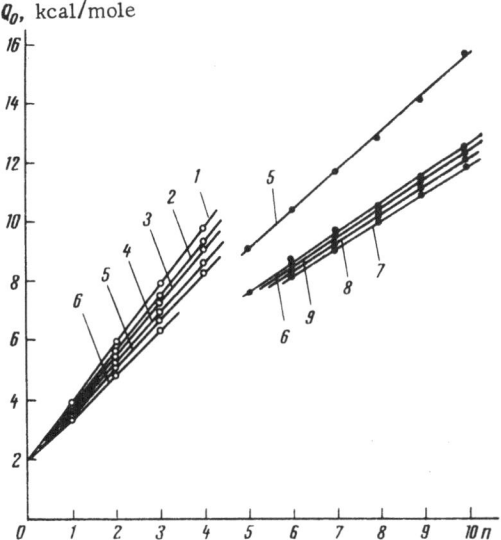

Fig. 40. Relation of Q_0 to n for the C_n hydrocarbons on silica gels of different pore sizes: 1) d = 32 Å; 2) d = 46 Å; 3) d = 70 Å; 4) d = 104 Å; 5) d = 140 Å; 6) d = 300 Å; 7) d = 1000 Å; 8) d = 710 Å; 9) d = 4100 Å.

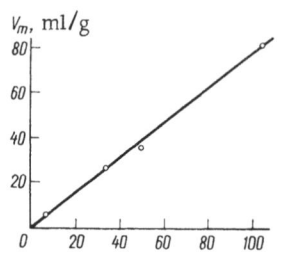

Fig. 41. Relation of V_m for n-hexane to s for macroporous silica gel at 100°C.

Figure 41 shows that a silica gel with sufficiently wide pores has (as for graphitized carbon, Fig. 7) V_m proportional to s, with $V_s = V_m/s$ almost independent of s [7, 15].

The Q_0 for normal alkanes on silica gel are linearly related to n (Fig. 40):

$$Q = a + bn, \qquad \text{(III, 2)}$$

with a and b tending to limits independent of s, n, and pore size as the last increases [15, 59].

The V_s for a given hydrocarbon are almost independent of s for silica gels with sufficiently large pores [15, 58, 59] (Fig. 42a), so these are physicochemical constants of the system at a given temperature. Reduction in pore size causes V_s, Q, a, and b to increase, especially when n is large, because of increase in adsorption energy as the pores become narrower (Fig. 42b).

If the specimen is small, the molecules are adsorbed preferentially on the best sites, and so V_m increases if the mean size of the pores is sufficiently reduced and s is increased [15, 59] (Fig. 43). The selectivity is also inversely related to pore size for silica gel (Fig. 44), but the separation criterion K_1 is almost constant (Fig. 45), and so it is better for the purposes of analysis to use silica gels of the larger pore sizes, since there is less peak broadening and much more rapid analysis while retaining a given K_1 [15, 57].

Figure 46 shows curves recorded on silica gel of fixed s but different mean pore size d [15, 57]. Increase in d by a factor 3 reduces the delay times by about a factor 4, while K_1 for the large d is somewhat greater, because the slow exchange for d small causes much more broadening.

In general, the retention is determined by the geometrical structure of the pores and by the chemical nature of the surface, as well as by the molecular weight, molecular geometry, and electronic structure. The column temperature also influences the re-

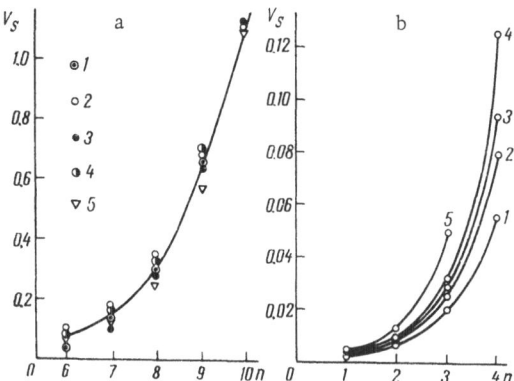

Fig. 42. Relation of V_s to n and pore size for normal hydrocarbons on silica gels of pore sizes: a: 1) 300 Å; 2) 710 Å; 3) 1000 Å; 4) 4100 Å; 5) mean data of [58]; b: 1) 140 Å; 2) 104 Å; 3) 70 Å; 4) 46 Å; 5) 32 Å.

sult. For each range of boiling points, given molecules similar in geometry and electronic structure, there is an optimal porosity giving relatively rapid separation with minimal broadening [15, 57]. For instance, silica gels for separating light gases should have d not more than 20 Å, whereas for light hydrocarbons (boiling points below 10°C) they should have d from 50 to 200 Å, and for reasonably rapid analysis of hydrocarbons of higher boiling point (and some derivatives of these) d should be even larger.

These conclusions are confirmed by results for porous glasses, which recently have come into use in gas chromatography [62, 63]. Grebenshchikov and Molchanova [64] first showed that alkali borosilicate glasses after appropriate heat treatment are readily attacked by acids and alkalis to give porous dentate structures. Zhdanov found that the properties of porous glasses are very much dependent on the composition and heat treatment of the initial material [29]; adjustment of these factors provides pore sizes from 8 to 10,000 Å, the range of pore size for a given specimen being narrow, which is especially important for gas chromatography.

The time of etching in HCl provides control of pore depth; bulk-porous and surface-porous materials can be produced, with the mean pore size controlled by the previous treatment. Surface-porous glass gives narrow peaks and rapid analysis [65] (Fig. 47).

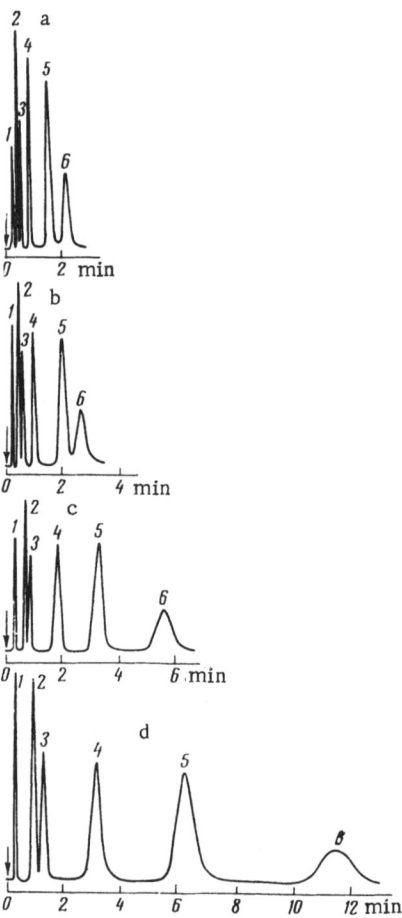

Fig. 43. Mixture of methane 1, ethane 2, ethylene 3, propane 4, propene 5, and butane 6 used
at 80°C with 150 × 0.45 cm columns of silica gels
of various pore sizes (hydrogen 50 ml/min, sample
volume 0.02 ml, flame-ionization detectors): a)
d = 104 Å; b) d = 70 Å; c) d = 46 Å; d) d = 32 Å.

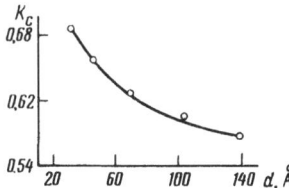

Fig. 44. Selectivity factor K_s for silica gel as a function of mean pore diameter (ethane and propane, 150 × 0.45 cm column, 25°C, carrier gas 50 ml/min).

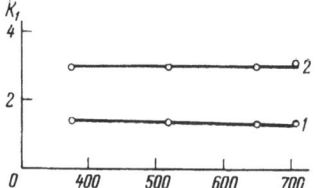

Fig. 45. Separation criterion K_1 as a function of s (m²/g) for: 1) ethane and ethylene; 2) propane and propene.

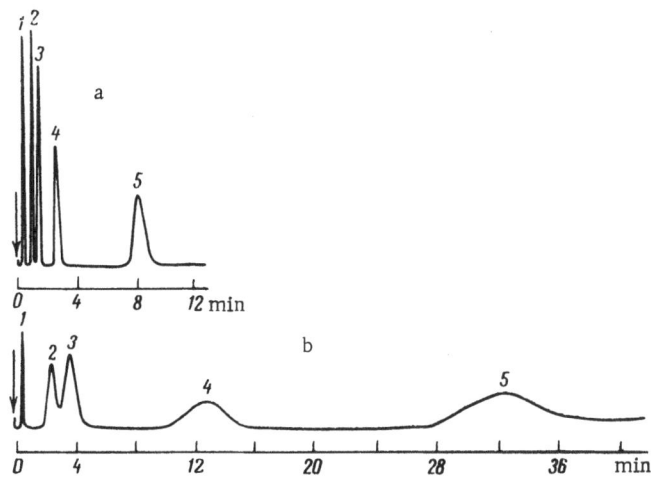

Fig. 46. Mixtures of methane 1, ethane 2, ethylene 3, propane 4, and propene 5 used at 50°C on two silica gels of equal s but different d (0.02 ml sample, hydrogen 50 ml/min); a) d = 70 Å, column 100 × 0.4 cm; b) d = 22 Å, column 58 × 0.4 cm.

The advantages of surface-porous capillary columns have already been mentioned above.

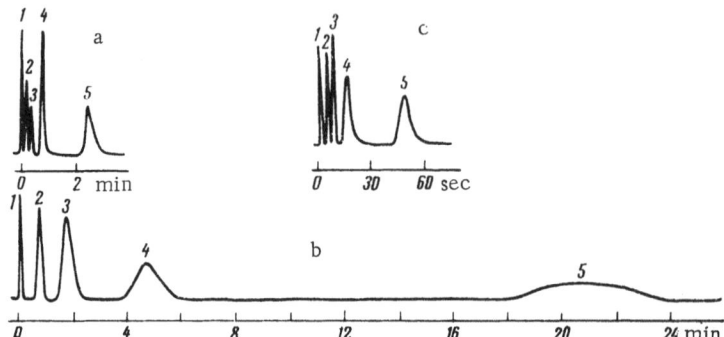

Fig. 47. Mixtures of methane (1), ethane (2), ethylene (3), propane (4), and propene (5) used with hydrogen carrier on 100 × 0.4 cm columns of porous glass: a) surface-porous glass, 68 ml/min; b) bulk-porous glass, 68 ml/min; c) surface-porous glass with high gas flow rate.

STRUCTURE OF ADSORBENT PORES
AND BAND BROADENING

The separating capacity of a column is determined by the selectivity of the adsorbent and by the peak broadening, the latter being the main factor interfering with sharp separation. Other things being equal, the column giving less broadening is the better.

There are various causes of broadening: kinetic, thermodynamic, and diffusion. The most important are as follows:

1. Broadening due to nonlinearity in the equilibrium isotherm (deviation from Henry's law): sharpening of the leading edge and broadening of the trailing edge or vice versa, since the velocity along the column varies with the concentration.

2. Broadening due to diffusion during motion along the column.

3. Broadening due to delay in adsorption and desorption.

4. Broadening related to sample injection.

The first cause may be eliminated by choosing an adsorbent with isotherms as linear as possible. One reason for the success of gas − liquid chromatography is linearity in the isotherms (adsorption at smooth liquid surfaces [66-68] and solution [69]) over wide ranges. Linear isotherms may be produced in gas adsorption (at high temperatures) by the use of nonporous or homogeneously macroporous adsorbents produced by modification (thermal, chemical, or geometrical). Initial linear parts are produced for light gases and hydrocarbons on homogeneously microporous adsorbents and porous crystals. The peak symmetry in such cases points to absence of broadening due to nonlinearity in the equilibrium isotherms.

Although it is at present at least clear how the first cause of broadening can be removed, it is quite impossible to suppress entirely the broadening due to diffusion; one can only minimize it.

The diffusion processes in a band on a column of granular adsorbent are very complex [70-75]. There is always molecular diffusion in the gas phase. In addition, the gas speed varies from point to point with the shape and packing of the granules, which is related to the wall effect, since the flow resistance near the wall is less than that of the center.

It is usually considered that the gas in such a column tends to swirl around the grains, which also causes some broadening, but the role of this eddy diffusion has been disputed [76].

A capillary column owes its performance mainly to the absence of the specific diffusion processes in granular materials; but an unfilled capillary also shows broadening due to the parabolic velocity profile [77].

Delay in adsorption or desorption causes broadening [74]. Delay in adsorption causes the material in the gas to advance (broadening of the leading edge), while delay in desorption causes broadening of the trailing edge. The rates of adsorption and desorption are unequal, so such broadening may be unsymmetrical.

Broadening due to finite mass-transfer rates is closely related to diffusion, since adsorption has three stages: motion up to the surface by external diffusion, internal diffusion in the pores and surface diffusion [78], and finally adsorption proper. The last is virtually instantaneous [79], and the desorption rate can be raised by heating the column, so the overall rate is governed by the rates of the first two stages and the converse stages.

These various causes can be evaluated only very crudely, on account of the variations in grain shape and size, variations in packing and porosity, and differences in the surfaces, including accessibility. These uncertainties prevent the use of kinetic theory.

The effects of external diffusion may be minimized by using an adsorbent of uniform grain size uniformly distributed in the column. The number of contacts of spheres poured freely into a column varies from 4 to 12 (packing ranging from very open to very close [80]), while irregular rough unequal grains pack even more unevenly. Hence a real column always has a certain velocity distribution.

Adsorption cannot occur until a molecule in the gas has approached the surface of a grain; the rate of external mass transfer is thus dependent on the mode of gas flow as well as on the kinetic energy of the molecule. In laminar flow, mass transfer occurs only by ordinary diffusion, whereas forced mixing occurs in turbulent flow. Reynolds' number for the usual flow rates in packed columns indicates that the flow is usually laminar; but in some rapid columns it is possible for the flow to be turbulent [81].

A study of any one of these factors requires minimization of the effects of the others [82]. Hence broadening should first be examined on the empty column. The following expression [83] gives the height H equivalent to a theoretical plate:

$$H = \frac{2D}{u} + \frac{1}{24} \frac{r^2 u}{D}, \qquad (\text{III, 3})$$

in which D is the coefficient of molecular diffusion, u is linear velocity, and r is radius. Figure 48a shows H as a function of u for an empty column with methane in hydrogen [83]; (III, 3) implies that the linear relation of Hu to u^2 can be used to find D.

The effects of swirling and uneven velocity distribution may be examined via the relation of H to u for columns of the same r filled with nonporous grains, with the gas carrying some unadsorbed substance. Figure 48b shows the relation of H to u (also for methane) for a column filled with 0.25-0.5 mm glass spheres. H increases with u somewhat at high u. In that case we can completely neglect the term in van Deemter's equation [84] related to the adsorption kinetics, since the glass spheres are nonporous, have low s, and almost completely fail to adsorb methane. The coefficient λ in that equation is then one. The D calculated for this case differs somewhat from that determined by other methods, but good agreement is obtained if the relation of H to u is given the

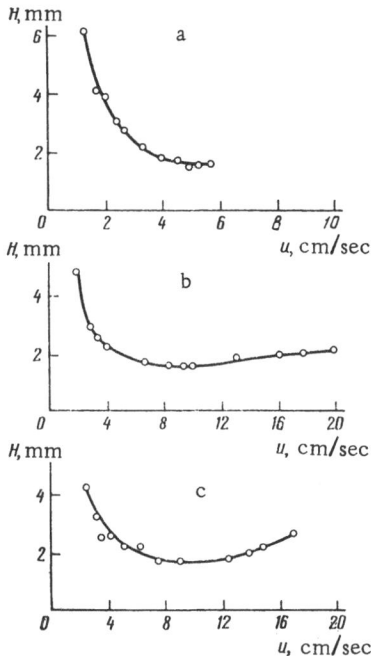

Fig. 48. Relation of H to u for methane in 300 × 0.4 cm columns (carrier gas hydrogen): a) empty column, 25°C; b) column filled with nonporous glass sphere, 25°C; c) column filled with macroporous spherical grains, 46°C.

form of [85]:

$$H \approx \frac{1.2D}{u} + \frac{1}{24}\frac{r^2 u}{D} \ . \ \text{(III, 4)}$$

H was almost as above in the range of optimal u for columns packed with NaCl grains of various sizes. The proportion of free space was directly related to the grain size and the spread in size, while the uniformity of packing (and hence the performance) deteriorated.

Figure 48c shows H−u results for methane (almost unadsorbed) used with macroporous spherical grains. In this case H rose at small u (reduced performance) on account of the broadening due to diffusion into the pores, where there is no general circulation [74].

The relation of H to u was examined for nonporous NaCl with the fairly strongly adsorbed n-hexane with the same sizes for adsorbent grains and column, in order to elucidate the adsorption kinetics and the role of external diffusion. Here the minimum H was even larger, and H began to rise at even lower u, on account of delay in exchange of molecules with the surface.

External and internal diffusion in the pores should increase in importance with the grain size if the compound is fairly strongly adsorbed. Figure 49 shows H against the bulk rate w for the carrier gas* for propane (unspecifically adsorbed) on silica gel (d = 80 Å) as spherical grains. The performance deteriorates considerably

* The bulk velocity w equals the linear velocity u multiplied by the effective cross-section of the column.

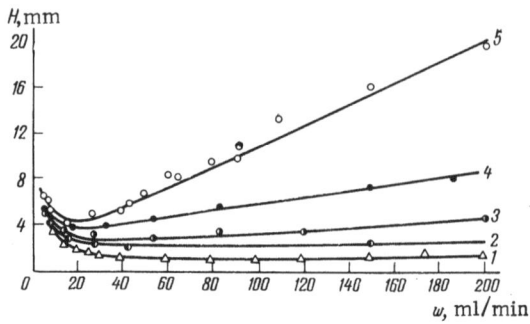

Fig. 49. Relation of H to w for propane at 60°C for silica
gel (d ≈ 80 Å) packed as spherical grains of size: 1) ~ 0.25
mm; 2) 0.13 − 0.25 mm; 3) 0.25 − 0.5 mm; 4) 0.5 − 1.0
mm; 5) 1.0 − 2.0 mm.

as the grain size increases; the region of w corresponding to low
H becomes narrow, and the slope of the right branch (here determined mainly by the adsorption kinetics) becomes pronounced.
Grain sizes of 0.25 mm or less gave a very low slope in the right
branch, so diffusion here contributes very little to the broadening.
Large grains (0.5-1.0 mm and, especially, 1.0-2.0 mm) produce
much broadening from internal and external diffusion.

The delay times for the C_1-C_4 saturated hydrocarbons, as
determined from the maxima under fixed conditions, remain independent of the size of the spherical grains of silica gel (Fig. 50),
but the half-widths of the peaks increase with the grain size. Coincidence of the peaks shows that the V_S are the same, so these
C_1-C_4 hydrocarbons are adsorbed on the entire accessible surface
even for grain sizes of 1.0-2.0 mm; but the increase in half-width
shows that there are increased delays in diffusion into and out of
the grains [82].

These diffusion effects should be diminished on raising T.
Figure 51 shows H−u curves for a column with 1.0-2.0 mm spherical grains of silica gel at different temperatures. Here the diffusion effects should be pronounced. The column becomes more
effective as T is raised; the region of low H becomes broader, and
the slope of the right branch is reduced. Hence diffusion is very
important for grains over 1 mm in size if T is too low.

The carrier gas also affects the diffusion under otherwise
fixed conditions, as Fig. 52 shows from the H−w relation for pro-

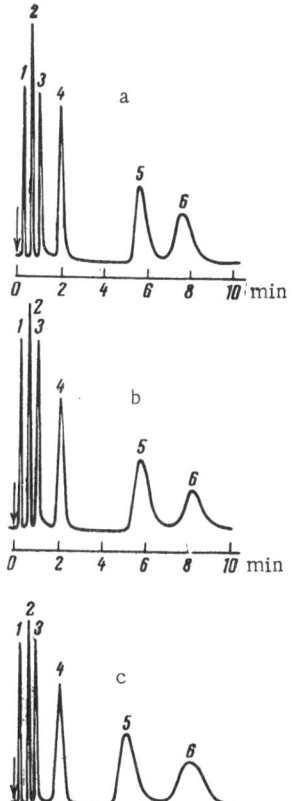

Fig. 50. Mixtures of methane (1), ethane (2), ethylene (3), propane (4), propene (5), and n-butane (6) at 50°C on microspheres of silica gel (hydrogen, 60 ml/min, column 100 × 0.45 cm, sample < 0.002 ml): a) < 0.25 mm; b) 0.25 − 0.5 mm; c) 0.5 − 1 mm.

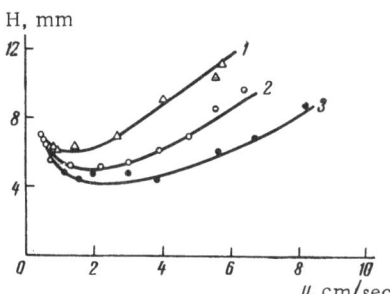

Fig. 51. Effects of column temperature on H − u relation for propane (100 × 0.4 cm column filled with 1.0-2.0 mm spheres of silica gel, d = 80 Å, helium): 1) 28°C; 2) 70°C; 3) 130°C.

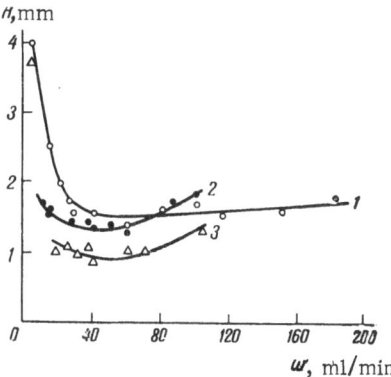

Fig. 52. Relation of H to w for propane at 50°C (column 100 × 0.4 cm, silica gel with d ≈ 100 Å as spherical grains 0.5-1.0 mm) with carrier gases: 1) He; 2) Ar; 3) CO_2.

pane on silica gel (d≈100 Å) as 0.25-0.5 mm spheres. The heavier gas (lower D) gives better performance, but the region of low H is much narrower, because the low D means appreciable delay at lower w.

Uniform adsorbent packing is important. Figure 49 shows that a very narrow fraction near 0.25 mm gives a higher efficiency

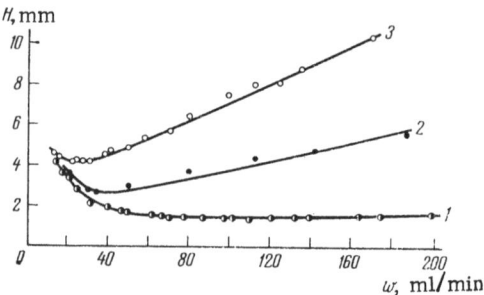

Fig. 53. Relation of H to w for methane at 30°C on 100 × 0.5 cm columns of activated charcoal of grain sizes: 1) 0.25-0.5 mm; 2) 0.5-0.1 mm; 3) 1.0-2.0 mm.

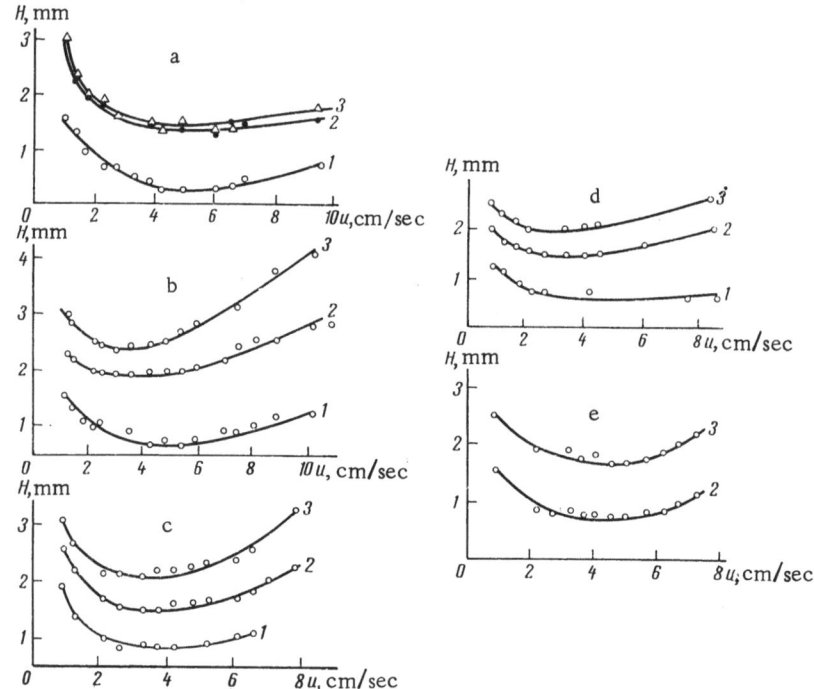

Fig. 54. Relation of H to u for: 1) oxygen; 2) nitrogen; 3) methane on zeolites at 22°C (carrier hydrogen, 2 ml gas samples): a) VNIINP* CaA, spherical; b) VNIINP CaA; c) Linde 5A; d) VNIINP CaX; e) Linde 13X.

*VNIINP is the standard of the All-Union Scientific Research Institute of the Petroleum Industry.

than a smaller mean size with a larger spread (0.13-0.25 mm fraction), on account of improved uniformity and thus reduced broadening from variations in gas speed.

Activated charcoal gives similar results; here again the performance is inversely related to size, on account of diffusion [82] (Fig. 53).

Figure 54 shows the relation of H to u for various zeolites and for substances differing in D. The optimal u for a 4 mm zeolite column is 3-6 cm/sec, the performance being less good for the heavier compounds (on a given column), with the region of minimum H narrowing and the slope of the right branch increasing.

The H−u relation (or H−w) reveals the effect of d in the range 15 to 150 Å for a fixed grain size of 0.25-0.5 mm under otherwise fixed conditions. Figure 49 indicates that broadening by external diffusion plays little part for this grain size, while Fig. 55 shows that the performance increases with d up to a certain limit, which is reached in the range 75-150 Å, where H is virtually independent of w for w > 30 ml/min. The range of low H becomes narrower at smaller d, and the slope of the right branch increases.

Equilibrium is attained in adsorption for these compounds at these temperatures, since V_s is independent of w even for the smallest d within certain limits, but the time actually spent by a molecule on the surface varies with the width and depth of the pores, and so the broadening tends to be inversely related to d, since the inhomogeneities of structure and field in the silica gel tend to increase as d becomes smaller.

The performance of a column is affected by the pore depth as well as by the mean pore diameter; Fig. 56 shows the H−u relation for the columns of Fig. 47, which had bulk-porous and surface-porous glass fillings. The high mass-transfer rate for surface-porous glass gives rise to lower H and a broader region of low H The resolving power in linear gas-adsorption chromatography is expressed as the ratio of the difference of retention times Δt_R (reckoned from peak positions, this being governed mainly by the Henry constants), to $\Delta t_1 + \Delta t_2$, the sum of the half-widths (governed by the diffusion and mass-transfer delays). In the general case, the width is governed by the shape and size of the grains, by the carrier gas, and (especially) by the adsorption and desorption ki-

netics, which may be dependent on the nature of the compound, on
the temperature, and on the nature and geometrical structure of
the adsorbent. Normal adsorbents (silica gel, activated charcoal,
etc.) have the pores passing right through the thickness of the grain,
so the internal diffusion paths are very long, which greatly retards
lengthwise diffusion in the pores, and which (under dynamic conditions) may lead to unequal use of the internal surface. Glasses
with only surface porosity allow one to use much longer analysis
times without loss of resolving power. Table 8 gives t_R for hydrocarbons on such columns, as well as K_1 (separation criterion), half-
width Δt, number N of theoretical plates, and height H equivalent
to one theoretical plate. Although surface-porous glass reduces
the analysis time by about a factor 8, bulk-porous glass gives higher
K_1 and N. Such columns allow the use of high u in order to reduce
the time even further, without loss of resolving power. Figure 47
shows a chromatogram for the mixture of Table 8 run on surface-
porous glass at $u \approx 20$ cm/sec, the t_R for propene here being only
~50 sec [62, 63, 65].

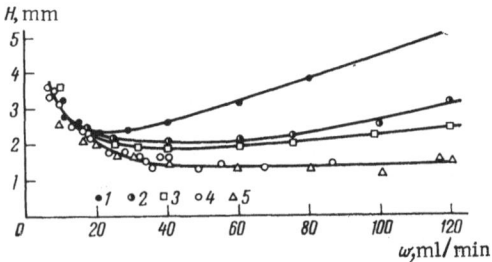

Fig. 55. Relation of H to w for propane at 50°C
on 0.25-0.5 mm silica gels (100 × 0.4 cm column,
helium) with d of: 1) 16 Å; 2) 23 Å; 3) 35 Å; 4) 75 Å.
5) 150 Å.

Table 8. Performance Parameters of Columns Filled

Glass grains	t_R, sec					Δt, sec				
	CH_4	C_2H_6	C_2H_4	C_3H_8	C_3H_6	CH_4	C_2H_6	C_2H_4	C_3H_8	C_3H_6
Bulk porous	13	56	117	295	1260	1.8	11.4	22.2	63	213
Surface porous	13	24	35	59	167	0.6	1.8	3.0	6.24	17.7

Broadening due to injection is as for gas−liquid chromatography and so will not be discussed here.

The performance of an adsorption column is determined by the pore size and depth, the grain size, the spread in grain size, the temperature, the carrier gas, and the compound.

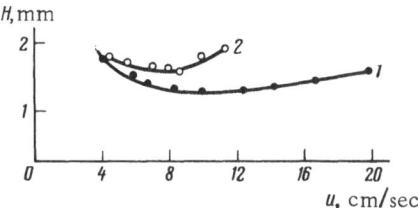

Fig. 56. H−u relation for propane at 50°C on columns 100 × 0.4 cm (hydrogen carrier) filled with: 1) surface-porous glass; 2) bulk-porous glass.

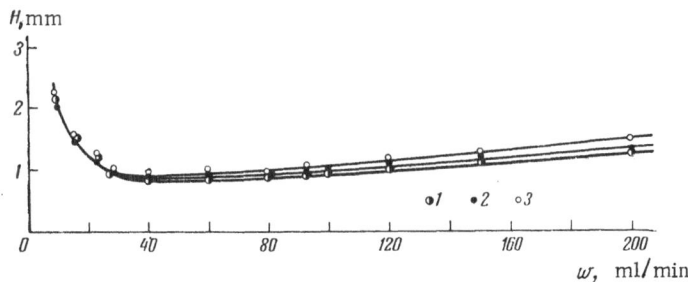

Fig. 57. H−w relations for normal alkanes at 100°C on macroporous silica gel (s = 30 m^2/g, nitrogen): 1) octane; 2) nonane; 3) decane.

with Bulk-Porous and Surface-Porous Glasses [62]

K_1				N				H, mm			
CH_4/C_2H_6	C_2H_6/C_2H_4	C_2H_4/C_3H_8	C_3H_8/C_3H_6	C_2H_6	C_2H_4	C_3H_8	C_3H_6	C_2H_6	C_2H_4	C_3H_8	C_3H_6
3.25	1.82	2.1	3.5	133	154	122	194	7.5	6.5	8.1	5.2
4.2	2.3	2.6	4.5	980	700	515	495	1.0	1.3	1.9	2.0

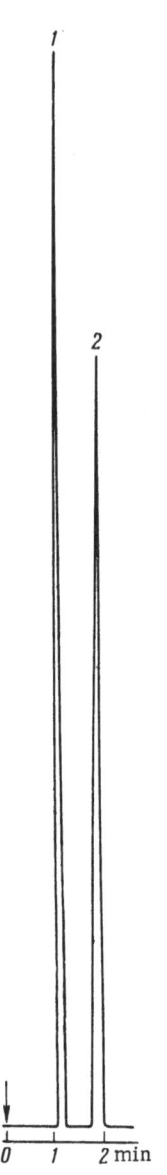

Fig. 58. Separation of pentane (1) and
hexane (2) at 100°C in a 100 × 0.4 cm
column containing macroporous silica
gel (s = 60 m²/g, 0.13-0.18 mm grains,
nitrogen 50 ml/min, performance
equivalent to 2000 theoretical plates.

An efficient column requires a macroporous adsorbent of uniform grain size, the best as regards flow resistance being ~0.2 mm. A macroporous adsorbent will give $H \approx 1$ mm or less (Fig. 57) even in grain sizes of 0.25-0.5 mm, or over 1000 theoretical plates per 1 m of column. Smaller and more uniform grain sizes give even higher performance, in that the broadening due to uneven velocity distribution is reduced; $H = 0.5$ mm or less are attainable, or several thousand theoretical plates per 1 m (Fig. 58).

IMMOBILE SUPPORTS

FOR GAS CHROMATOG-

RAPHY

In gas—liquid chromatography, a thin film of a nonvolatile solvent (the immobile phase) is supported on a solid carrier, which should maintain the liquid as a uniform thin film. In gas-adsorption chromatography a solid support may be filled with a solid adsorbent [6-8].

The solid support must meet the following requirements:

1. It must be free from micropores, while its macropores must communicate via large holes.

2. The specific surface should be small (0.5-10 m²/g).

3. It should be chemically inert.

4. It should adsorb weakly and unspecifically.

5. It should be stable at least up to 350°C; and

6. The grains should be as equal as possible in size (~0.2 mm), of simple smooth shape, and of adequate mechanical strength.

So far there is no ideal support meeting all these requirements. Various values have been quoted for the s at which separations are unaffected; Kaiser [86] gives ~3 m^2/g, while Ettre [87] gives about 1 m^2/g.

The large s of ordinary porous adsorbents lead to strong adsorption, so preference for supports is given to natural diatomaceous earths, which have low s. The skeleton here consists mainly of hydrated amorphous and fairly fine-pored silica penetrated by much larger holes, which (after heat treatment) determined the macroporosity. Many diatomites of identical structure have been reported; the macropores are usually ~1 μ in size. Diatomites contain traces of various oxides: ~90% SiO_2, ~4% Al_2O_3, ~1.5% Fe_2O_3, ~0.2% TiO_2, ~0.5% CaO, ~0.5% MgO.

Ottenstein [88] divides diatom-earth supports into two types. To the first type he assigns supports prepared from natural diatomite, ground, pressed, and fired at 900°C. This includes all supports made from diatomaceous refractory brick-earths: C-22, INZ-600, etc. To the second type he assigns supports made by sintering natural diatomite with fluxes (e.g., sodium carbonate) about 900°C, e.g., Celite-545 and chromosorb types W, G, and A.

Diatomite supports may additionally be fired at 1100 and 1350°C [89-91], which substantially reduces s (by sintering of the smaller pores) and so greatly reduces the uptake in adsorption.

These forms of silica contain about 10% of other minerals, which may enter the skeleton or may form a surface contamination. The surface of silica is (see Chapter II) covered with silanol [I] or siloxane [II] groups:

$$\left(-\underset{\vert}{\overset{\vert}{Si}}-OH\right) \qquad \left(-\underset{\vert}{\overset{\vert}{Si}}-O-\underset{\vert}{\overset{\vert}{Si}}-\right).$$
$$\qquad\qquad I \qquad\qquad\qquad\qquad II$$

It has been concluded [92] that these diatomite supports are essentially silica gels of low s, but the presence of alumina would

Table 9. Some Supports for Gas Chromatography

Support	Source	s, m²/g	Pore volume, cm³/g	Density, g/cm³
Celite-545 . . .	⎫	~1		
Chromosorb A . .	⎪	4.0	—	0.38
Chromosorb P . .	⎬ Johns Manville Corp.	1.0	—	0.18
Chromosorb Q . .	⎪ (USA)	0.5	—	0.47
Chromosorb W . .	⎪	2.7	—	
Chromosorb T . .	⎭	7—8	—	
Sterhamol . . .	Sterhamol Werke (Ger.)	9.7	0.09	0.44
Risorb	VUOS (Czech.)	2—6.8	0.013	0.50

imply that the surface has aprotonic acid centers as well as protonic ones [93], the former being strong electron acceptors and giving rise to surface charge-transfer compounds, as well as catalytic activity.

Ottenstein's two types differ in pH of water extracts. The hydroxylated surface of amorphous silica is weakly acid [94], while the alkaline response of the second type is due to the use of sodium carbonate as a flux. The first type gives pH 6.0-6.5, whereas the second gives pH 8-8.5.

James and Martin [95] proposed Celite-545 as an inert support; this is a diatomaceous material used in removing heavy fractions in petroleum processing and also as a white filler for paper. Various materials used for other purposes have also been proposed. The Johns Manville Corp. has developed various supports especially for gas chromatography, such as Chromosorb types A, P, W, T, and Q [96]. Chromosorb Q is mechanically stronger than Celite-545 and is also twice as dense; it is recommended for use with up to 5% liquid phase. The surface has been modified by attachment of alkylsilyl groups. Chromosorb A [96] is meant particularly for preparative purposes and can carry up to 20-30% of liquid.

Table 9 gives the characteristics of the commoner solid supports used in gas chromatography.

The particles of diatomite in Ottenstein's first type fuse together on firing [97] and some of the amorphous silica becomes crystalline cristobalite; the small pores are sealed and s is much reduced. The traces of metals form silicates or complex oxides.

The material becomes white, since the iron oxide is converted to a colorless iron silicate.

Baker et al. [98] have tried to correlate the structure of the support with the distribution of liquid on the surface; they measured [98] s and the pore distribution. The chemical composition, acidity, and crystal structure have also been examined [99]. Supports of the first type have pores 0.4 to 2 μ in size (mean ~1 μ), while those of the second type have much large pores (8-9 μm

A small quantity of liquid enters mainly the small pores, as is clear from the reduction in s [98]; the smoother parts of the large pores may remain largely uncovered. When there is 15% liquid, about 12% goes to the small pores and only 2-4% to the rest of the surface [100, 101].

Diatomite supports show fairly pronounced chemisorption and specific adsorption; various methods are used to suppress these. It is usually assumed that the liquid phase (especially if polar) produces some deactivation, but it has been shown [102, 103] that the surface is not deactivated even by strongly polar liquids, which indicates [102, 103] that deactivation occurs when the liquid forms a hydrogen bond to the hydroxylated surface. Usually a weakly polar liquid does not produce deactivation. In that case, small amounts of polar liquid (often surface-active) are added [85].

Sometimes the carrier is treated to produce symmetrical peaks by saturating the carrier gas with a volatile polar compound, in particular water in the separation of alcohols [104], formic acid in the separation of fatty acids [105], ammonia for amines [106], etc. These deactivator compounds are not recorded by the flame-ionization detector. It has been suggested that the support should be treated with alkali in the analysis of compounds containing nitrogen, especially amines [107, 108]. Acid washing removes some mineral impurities and improves the separation [109, 110], but it is hazardous in some separations, because it causes reactions with the compounds [111].

Chemical modification is the most effective way of deactivating a surface; silane derivatives were first used for this purpose in 1958 [104, 114] by analogy with the modification of adsorbents and fillers [112, 113]. Bohemen et al. [115] used hexamethyldisilazane to modify the surface of the carrier; various ways of using this were subsequently [116-118] proposed. The best success with a modified carrier was in the separation of traces of ethanol from methanol [119]. The $-O-Si-OR_3$ bond is stable up to 360°C [120].

Chemical modification greatly reduces the specificity but does not produce complete surface inertness, on account of the steric hindrance to modification discussed in Chapter II, which prevents the formation of a very dense layer of inert groups attached to the surface (p. 53).

Reactions sometimes occur during separation under the influence of the support; terpenes isomerize on an acid support [121], e.g., pinene becomes camphene [122]. Carriers of the first type often cause dehydration of alcohols [123].

Supports are also deactivated by coating them with solids; in particular, diatomite supports are coated with silver up to 40% by weight, which gives rise to symmetrical peaks [124]. However, this support cannot be used for analysis of some compounds containing sulfur and nitrogen. Supports may also be coated with poly-tetrafluoroethylene (PTFE), which considerably reduces the activity [125, 126], though the performance of the column is then rather poor. PTFE of grade F-42P, which is soluble in acetone, has been deposited to the content of 7% on diatomite [176]. A column 160 × 0.3 cm gave good separation of the hydrocarbons up to C_{31}, of methyl esters of synthetic fatty acids, of the C_5-C_{21} cyanides, and of normal and branched-chain alcohols.

If solution in the liquid is the main cause of retention in equilibrium gas-liquid chromatography, the volumes V_R should [127] be linearly related to the volume of liquid phase:

$$V_R = V_g + v K, \tag{III, 5}$$

in which V_g is the retained volume of unsorbed gas, K is the partition coefficient, and v is the volume of liquid. However, (III, 5) often does not apply [128], and the V_R of some substances even increase as v is reduced, because the carrier itself acts in adsorption when v is small (up to 5%).

A high liquid content produces poor separation because of slow diffusion in the liquid, especially at high gas-flow rates; it has been considered [98, 129] that the optimum proportion of liquid is therefore 15% of the weight of the support. The separation deteriorates above 15%, though the effect is dependent on the pore structure, on s, and on the uptake in adsorption (Fig. 59).

Active adsorbents have been proposed as carriers for volatile liquids, e.g., activated charcoal with 20% nitrobenzene for separating low-boiling hydrocarbons [130], or active alumina with 4.5% water for separating ethane, ethylene, acetylene, propane, and propene in coking gas [131]; but these mixtures are more conveniently separated by gas adsorption (Chapter V). Alumina fired at 400°C and bearing 7.5% squalane provides excellent separation of the alkanes up to butane [132]. TZK (tripolite from Zikeevsk quarry) has been used with oil coating to separate hydrocarbon mixtures [133].

Other supports are fluorinated hydrocarbons, especially PTFE (Teflon), and glass spheres. Teflon consists of crystalline blocks and thin fibers. A disadvantage of Teflon columns is their relatively poor performance, which is due to uneven distribution of the liquid between the fibers. Teflon has been used in the analysis of aqueous mixtures [134-136], water and formaldehyde [137-139], and fatty acids [140], in particular mixtures of formic acid with acetic acid [141].

Glass spheres are more active in adsorption than Teflon [142], because the glass adsorbs specifically; but s is very small, so the spheres hold only a very small volume of liquid phase. However, the absence of pores means that glass spheres provide effective columns [143], which allow relatively low temperatures to be used in separating high-boiling substances. In particular, glass spheres with 0.05% liquid have been used with columns operated 250°C below the boiling points of components of the mixture [144, 145].

Other supports have been detergents [146-148], NaCl [149-152], sand [153], polyethylene [154], metal spirals [155-157], unglazed porcelain [99, 158], carborundum [159, 160], sintered glass flour [161], macroporous glass [162], macroporous silica gel [163, 164], marine sponges [165], and vermiculite [166].

Macroporous silica gels of low s have advantages over diatomite carriers: high chemical purity, homogeneity, mechanical strength. Dehydroxylated macroporous silica gel produced by hy-

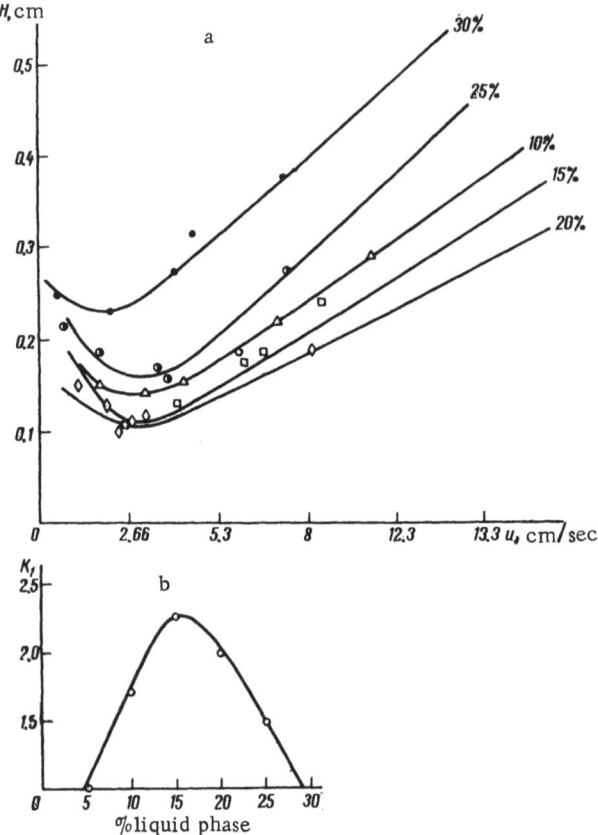

Fig. 59. a) H—u relation for benzene for a 100 × 0.4 cm column filled with diatomaceous support bearing various percentages of squalane. b) Relation of K_1 (benzene and toluene) to percent liquid phase.

drothermal treatment [42-46] has been used as a support. Firing of this at 900°C greatly improved the behavior, because the tendency to specific adsorption was reduced (Chapter II).

Dehydroxylated macroporous silica gel bearing squalane gave better results than Celite-545 bearing squalane in the separation of alcohols; the peaks were more symmetrical and better separated.

Graphitized carbon black has also been used [167] at high temperature; it differs from other materials in needing no special treatment. Used with a nonpolar liquid phase, it gives symmetrical peaks for polar substances (alcohols, ketones, amines). The sole disadvantage is the poor mechanical strength of the granules, so care is needed in filling the column.

Martin [67, 68] showed that adsorption on the liquid plays a considerable part in gas – liquid chromatography, especially when substances of low solubility are separated on a polar liquid.

Although it is often believed that the solubility isotherms are linear up to high concentrations, it has been shown [168] that they are often convex to the concentration axis. Consequently [169] the peaks in gas−liquid chromatography are often unsymmetrical, and the retention times vary with the size of the sample. This curvature is additional to the curvature caused by adsorption on the inhomogeneous surface of the support and so reduces the performance of the gas – liquid column.

This all shows that supports resemble adsorbents in standing in need of considerable improvement. Nonporous supports with nonspecifically adsorbing surfaces should be used, especially graphitized carbon blacks of low s. Optimal geometrical modification of silica gel should provide a homogeneous macroporous structure of low s with high pore volume; the macropores should be open and communicate via wide holes. The chemical and other features of such a surface are readily adjusted via chemical modification (Chapter II).

LITERATURE CITED

1.   C. Cercy and F. Botter, Bull. Soc. Chim. France, No. 11:3383 (1965).
2.   M. Mohnke and W. Saffert, in collection: Gas Chromatography 1962, M. van Swaay (ed.), London, 1962, p. 214.
3.   A. V. Kiselev, Zh. Fiz. Khim., 23:452 (1949).

4. A. V. Kiselev, in collection: Methods of Structure Examination for Finely Divided and Porous Bodies, Issue 1, Moscow, Izd. Akad. Nauk SSSR, 1953, p. 86.

5. A. V. Kiselev, in collection: Methods of Structure Examination for Finely Divided and Porous Bodies, Issue 2, Moscow, Izd. Akad. Nauk SSSR, 1958, p. 47.

6. E. Cremer, Angew. Chem., 72:512 (1959).

7. A. V. Kiselev, E. A. Paskonova, R. S. Petrova, and K. D. Shcherbakova, Zh. Fiz. Khim., 38:161 (1964).

8. A. V. Kiselev, I. A. Migunova, and Ya. I. Yashin, in collection: Gas Chromatography, Issue 6, Moscow, Izd. NIITÉKhim, 1967, p. 84.

9. S. Ross, J. K. Saelens, and J. P. Olivier, J. Phys. Chem., 66:696 (1962).

10. L. D. Belyakova, A. V. Kiselev, and N. V. Kovaleva, Anal. Chem., 36:1517 (1964).

11. R. L. Gale and R. A. Beebe, J. Phys. Chem., 68:555 (1964).

12. A. V. Kiselev, in collection: The Structure and Properties of Porous Materials, D. H. Everett and F. Stone (eds.), London, 1958, p. 195.

13. A. A. Isirikyan and A. V. Kiselev, J. Phys. Chem., 65:601 (1961).

14. I. Yu. Babkin and A. V. Kiselev, Zh. Fiz. Khim., 37:228 (1963).

15. A. V. Kiselev, Yu. S. Nikitin, R. S. Petrova, K. D. Shcherbakova, and Ya. I. Yashin, Anal. Chem., 36:1526 (1964).

16. P. C. Carman and F. A. Raal, Proc. Roy. Soc., A209:915 (1951).

17. B. G. Aristov, A. P. Karnaukhov, and A. V. Kiselev, Zh. Fiz. Khim., 36:2486 (1962).

18. B. G. Aristov, V. Ya. Davydov, A. P. Karnaukhov, and A. V. Kiselev, Zh. Fiz. Khim., 36:2758 (1962).

19. R. Venable and W. H. Wade, J. Phys. Chem., 69:1395 (1965).

20. A. V. Kiselev, Dokl. Akad. Nauk SSSR, 98:431 (1954).

21. A. V. Kiselev, E. A. Leont'ev, V. M. Luk'yanovich, and Yu. S. Nikitin, Zh. Fiz. Khim., 30:2149 (1956).

22. A. V. Kiselev, in collection: Proceedings by the Second International Congress on Surface Activity, Vol. 2, J. H. Schulman (ed.), London, 1957, p. 189.

23. C. R. Adams and H. H. Voge, J. Phys. Chem., 61:722 (1957).

24. W. G. Schlaffer, C. R. Adams, and J. N. Wilson, J. Phys. Chem., 69:1530 (1965).

25. A. V. Kiselev, Yu. S. Nikitin, and É. B. Oganesyan, Kolloidn. Zh., 28:662 (1966).

26. G. K. Boreskov, M. S. Borisova, O. M. Dzhigit, V. A. Dzis'ko, V. P. Dreving, A. V. Kiselev, and O. A. Likhacheva, Zh. Fiz. Khim., 22:603 (1948).

27. G. K. Boreskov, M. S. Borisova, V. A. Dzis'ko, A. V. Kiselev, O. A. Likhacheva, and T. N. Morokhovets, Dokl. Akad. Nauk SSSR, 62:649 (1948).

28. S. J. Gregg, The Surface Chemistry of Solids, London, Chapman and Hall (eds.), 1961; S. J. Gregg and K. S. W. Sing, Adsorption, Surface Area and Porosity, London, Academic Press, 1967.

29. S. P. Zhdanov, in collection: Methods of Structure Examination for Finely Divided and Porous Bodies, Issue 2, Moscow, Izd. Akad. Nauk SSSR, 1958, p. 117.

30. M. M. Dubinin, in collection: Chemistry and Physics of Carbon, P. L. Walker (ed.), New York, Dekker, 1966, p. 51.

31. J. R. Dacey and J. A. Fendley, in collection: The Structure and Properties of the Porous Materials, D. H. Everett and F. Stone (ed.), London, 1958, p. 142.

32. R. M. Barrer, in collection: The Structure and Properties of Porous Materials, D. H. Everett and F. Stone (eds.), London, 1958, p. 6.

33. R. M. Barrer, in collection: Non-Stoichiometric Compounds, L. Mondelcorn (ed.), New York, Academic Press, 1963, p. 309.

34. A. V. Kiselev and A. A. Lopatkin, Kinetika i Kataliz, 4:786 (1963).

35. G. J. Minkoff and R. H. E. Duffett, British Petroleum Magazine, No. 13, 1964, p. 16.

36. R. M. Barrer, Brit. Chem. Eng., No. 5, 1959, p. 1.

37. N. Brenner, E. Cieplinski, L. S. Ettre, and V. J. Coates, J. Chromatog., 3:230 (1960).

38. C. F. Spencer, J. Chromatog., 11:108 (1963).

39. R. A. van Nordstrand, W. E. Kreger, and H. E. Ries, J. Phys. Coll. Chem., 55:621 (1951).

40. H. E. Ries, Advan. Catalysis, 4:87 (1952).

41. N. V. Akshinskaya, A. V. Kiselev, and Yu. S. Nikitin, Zh. Fiz. Khim., 37:927 (1963).

42. N. V. Akshinskaya, V. E. Beznogova, A. V. Kiselev, and Yu. S. Nikitin, Zh. Fiz. Khim., 36:2277 (1962).

43. A. V. Kiselev, in collection: Gas Chromatography 1962, M. van Swaay (ed.), London, 1962, p. 34.

44. N. V. Akshinskaya, V. Ya. Davydov, L. T. Zhuravlev, G. Curthoys, A. V. Kiselev, B. V. Kuznetsov, Yu. S. Nikitin, and V. V. Rybina, Kolloidn. Zh., 26:529 (1964).

45. N. V. Akshinskaya, T. A. Baigubekova, A. V. Kiselev, and Yu. S. Nikitin, Kolloidn. Zh., 28:164 (1966).

46. N. V. Akshinskaya, V. Ya. Davydov, A. V. Kiselev, and Yu. S. Nikitin, Kolloidn. Zh., 28:3 (1966).

47. A. V. Kiselev and G. G. Muttik, Kolloidn. Zh., 19:562 (1957).

48. N. M. Kamakin and Ya. V. Mirskii, in collection: Methods of Structure Examination for Finely Divided and Porous Bodies, Issue 2, Moscow, Izd. Akad. Nauk SSSR, 1958, p. 190.

49. G. V. Vinogradov and L. V. Titkova, Kolloidn. Zh., 27:138 (1965).

50. G. V. Vinogradov, L. V. Titkova, N. V. Akshinskaya, N. K. Bebris, A. V. Kiselev, and Yu. S. Nikitin, Zh. Fiz. Khim., 40:886 (1966).

51. O. L. Hollis, Anal. Chem., 38:309 (1966).

52. A. V. Kiselev, Usp. Khim., 25:705 (1956).

53. A. V. Kiselev, in collection: Proceedings of the Second International Congress on Surface Activity, Vol. 2, J. H. Schulman (ed.), London, 1957, p. 179.

54. A. V. Kiselev and Yu. A. Él'tekov, Zh. Fiz. Khim., 31:250 (1957); A. V. Kiselev and Yu. A. Él'tekov, in collection: Proceedings of the Second International Congress on Surface Activity, Vol. 2, J. H. Schulman (ed.), London, 1957, p. 228.

55.   A. A. Isirikyan and A. V. Kiselev, Zh. Fiz. Khim., 31:2127 (1957).
56.   D. A. Vyakhirev, N. P. Chernyaev, and A. I. Bruk, Zh. Fiz. Khim., 34:1096 (1960).
57.   A. V. Kiselev and Ya. I. Yashin, Neftekhimiya, 4:494 (1964).
58.   A. V. Kiselev, R. S. Petrova, and K. D. Shcherbakova, Kinetika i Kataliz, 5:526 (1964).
59.   A. V. Kiselev and Ya. I. Yashin, Neftekhimiya, 4:634 (1964).
60.   D. P. Poshkus, Zh. Fiz. Khim., 39:2962 (1965).
61.   L. D. Belyakova, A. V. Kiselev and N. V. Kovaleva, Zh. Fiz. Khim., 40: 1494 (1966).
62.   S. P. Zhdanov, A. V. Kiselev, and Ya. I. Yashin, Zh. Fiz. Khim., 37:1432 (1963); in collection: Gas Chromatography, Moscow, Izd. NIITÉKhim, 1964, p. 5.
63.   Ya. I. Yashin, S. P. Zhdanov, and A. V. Kiselev, in collection: Gas Chromatographie 1963, H.-P. Angele and H. G. Struppe (eds.), Berlin, Akademie Verlag, p. 402.
64.   I. V. Grebenschikov and O. S. Molchanova, Zh. Obshch. Khim., 12:588 (1942).
65.   S. P. Zhdanov, A. V. Kiselev, and Ya. I. Yashin, Zh. Fiz. Khim., 37:1432 (1963).
66.   M. A. Khan, in collection: Gas Chromatography, 1962, M. van Swaay (ed.), London, 1962, p. 3.
67.   R. L. Martin, Anal. Chem., 33:347 (1961).
68.   R. L. Martin, Anal. Chem., 35:116 (1963).
69.   A. T. James and A. J. P. Martin, Biochem. J., 50:679 (1952).
70.   A. Keulemans, Gas Chromatography [Russian translation], Moscow, IL, 1959 [English edition: Second edition, New York, Reinhold, 1959].
71.   G. Schay, Theoretical Principles of Gas Chromatography [Russian translation], Moscow, IL, 1963.
72.   D. P. Timofeev, Adsorption Kinetics, Moscow, Izd. Akad. Nauk SSSR, 1962.
73.   N. M. Turkel'taub, L. N. Ryabchùk, S. N. Morozova, and A. A. Zhukhovitskii, Zh. Fiz. Khim., 19:133 (1964).
74.   A. A. Zhukhovitskii and N. M. Turkel'taub, Gas Chromatography, Moscow, Gostoptekhizdat, 1962.
75.   G. Cuiochon, Bull. Soc. Chim. France, No. 6, 1965, p. 3367.
76.   G. C. Giddings, Anal. Chem., 35:1338 (1963).
77.   M. Golay, in collection: Gas Chromatography, Proceedings of the Second International Symposium in Amsterdam and of the Conference on the Analysis of Mixtures of Volatile Substances in New York [Russian translation], Moscow, IL, 1961, p. 39.
78.   J. R. Dacey, Ind. Eng. Chem., 57:26 (1965).
79.   J. de Boer, The Dynamical Character of Adsorption [Russian translation], Moscow, IL, 1962, p. 49 [English edition: Oxford University Press, 1953].
80.   W. Smith, P. Foote, and P. Busony, Phys. Rev., 34:1271 (1929).
81.   V. Pretorius and T. W. Smuts, Anal. Chem., 38:274 (1966).

82. A. V. Kiselev and Ya. I. Yashin, Gas Chromatography, Transactions of the Third All-Union Conference on Gas Chromatography, Dzerzhinsk, Izd. Dzerzhinsk. Fil. OKBA, 1966, p. 131.
83. C. Giddings and C. F. Spencer, J. Chem. Phys., 33:1579 (1960).
84. J. J. van Deemter, F. J. Zuiderweg, and A. Klinkenberg, Chem. Eng. Sci., 5:271 (1956).
85. R. Kieselbach, Anal. Chem., 33:23 (1961).
86. R. Kaiser, in: Gas Chromatographie, Leipzig, Akademie Verlag, 1960, p. 39.
87. L. Ettre, J. Chromatog., 4:166 (1960).
88. D. M. Ottenstein, in collection: Advances in Chromatography, J. C. Giddings and R. A. Keller (eds.), New York, Dekker, 3:137 (1966).
89. A. B. Littlewood, J. Gas Chromatog., 1:20 (1963).
90. V. G. Berezkin, V. P. Pakhomov, L. L. Starobinets, and L. G. Berezkina, Neftekhimiya, 5:438 (1965).
91. R. I. Sidorov and A. A. Khvostikova, Zh. Fiz. Khim., 20:898 (1965).
92. R. H. Perrett and J. H. Purnell, J. Chromatog., 7:455 (1962).
93. A. V. Kiselev and A. V. Uvarov, Surface Sci., 6:401 (1967).
94. A. Ailer, Colloid Chemistry of Silica and Silicates, Moscow, Gosstroiizdat, 1959.
95. A. T. James and A. J. Martin, Biochem. J., 50:679 (1952).
96. Chromosorb G. T. A., FF-121, DD-124, DD-133, Johns Manville Corp.
97. D. M. Ottenstein, J. Gas Chromatog., 1:11 (1963).
98. W. J. Baker, E. W. Lee, and R. F. Wall, Gas Chromatography, H. Noebels, R. F. Wall, and N. Brenner (eds.), New York, Academic Press, 1961, p. 21.
99. G. Blandenet and J. Robin, J. Gas. Chromatog., 2:225 (1964).
100. J. C. Giddings, Anal. Chem., 34:458 (1962).
101. R. A. Keller and G. H. Stewart, Anal. Chem., 34:1834 (1962).
102. R. G. Scholz and W. W. Brandt, in: Gas Chromatography, N. Brenner, J. E. Callen, and M. D. Weiss (eds.), New York, Academic Press, 1962, p. 7.
103. V. Kusy, Anal. Chem., 37:1748 (1965).
104. H. S. Knight, Anal. Chem., 30:2030 (1958).
105. P. G. Ackman and R. D. Burgher, Anal. Chem., 35:647 (1963).
106. H. A. Saroff, A. Karmen, and J. W. Healy, J. Chromatog., 9:122 (1962).
107. J. F. O'Donnell and C. K. Mann, Anal. Chem., 36:2097 (1964).
108. J. L. Sze, M. L. Borke, and O. M. Ottenstein, Anal. Chem., 35:240 (1963).
109. A. Liberti, in collection: Gas Chromatography 1958, D. H. Desty (ed.), New York, Academic Press, 1958, p. 214.
110. E. C. Horning, K. C. Maddock, K. J. Anthony, and W. J. A. Vandenheuvel, Anal. Chem., 35:526 (1963).
111. J. Kallen and E. Hellbronner, Helv. Chim. Acta, 43:489 (1960); W. J. Zubyk and A. Z. Conner, Anal. Chem., 32:912 (1960).
112. I. Yu. Babkin, V. S. Vasil'eva, I. V. Drogaleva, A. V. Kiselev, A. Ya. Korolev, and K. D. Shcherbakova, Dokl. Akad. Nauk SSSR, 129:131 (1959).
113. I. Yu. Babkin and A. V. Kiselev, Zh. Fiz. Khim., 36:2448 (1962).
114. A. Kwantes and G. W. A. Rijnders, in: Gas Chromatography 1958, D. H. Destry (ed.), New York, Academic Press, 1958, p. 125.

115.   J. Bohemen, S. H. Langer, R. H. Perrett, and J. H. Purnell, J. Chem. Soc.,
        1960, p. 2444.
116.   P. Urone and J. F. Parcher, J. Gas Chromatog., 3:35 (1965).
117.   A. D. Atkinson and G. A. P. Tuey, Nature, 199:482 (1963).
118.   R. A. Dewar and V. S. Maier, J. Chromatog., 11:295 (1963).
119.   K. J. Bombaugh and W. E. Thomasen, Anal. Chem., 35:1452 (1963).
120.   T. B. Gavrilova, M. Krejci, H. Dubsky, and J. Janak, in collection: Czech.
        Chem. Commun., 29:2753 (1964).
121.   H. P. Burchfield and E. E. Storrs, in: Biochemical Applications of Gas Chromato
        graphy, New York, Academic Press, 1962, p. 468.
122.   A. L. Conner, in collection: Gas Chromatography 1958, D. H. Desty (ed.),
        New York, Academic Press, 1958, p. 214.
123.   A. V. Holmgren, in collection: Gas Chromatography, V. J. Coates, H. J.
        Noebels, and I. S. Fogerson (eds.), New York, Academic Press, 1958, p. 39.
124.   E. C. Omerod and R. P. W. Scott, J. Chromatog., 2:65 (1959).
125.   T. Onaka and T. Okamoto, Chem. Pharm. Bull. (Japan), 10:757 (1962).
126.   J. J. Kirkland, in collection: Gas Chromatography, L. Fowler (ed.), New
        York, Academic Press, 1963, p. 77.
127.   A. Keulemans, Gas Chromatography [Russian translation], Moscow, IL, 1959
        [English edition: Second, New York, Reinhold, 1959].
128.   T. Fukuda, Japan Analyst, 8:627 (1959).
129.   J. D. Cheshire and R. P. W. Scott, J. Inst. Petrol., 44:74 (1958).
130.   W. E. Tabconer and J. H. Knox, J. Chem. Soc., 1961, p. 782.
131.   F. Gebert and G. Herbst, Brennstoff-Chem., 43:20 (1962).
132.   J. Halasz, Z. Anal. Chem., 181:382 (1961).
133.   A. I. Tarasov, N. A. Kudryavtseva, A. V. Iogansen, and N. I. Lulova, in col-
        lection: Gas Chromatography, Transactions of the First All-Union Conference
        on Gas Chromatography, Moscow, Izd. Akad. Nauk SSSR, 1960, p. 280.
134.   O. F. Bennett, Anal. Chem., 36:684 (1964).
135.   R. Aubeou, L. Chompeix, and J. Reiss, J. Gas Chromatog., 16:7 (1964).
136.   W. M. Schwecke and J. H. Nelson, Anal. Chem., 36:689 (1965).
137.   K. J. Bombaugh and W. C. Bull, Anal. Chem., 34:1237 (1962).
138.   S. Sondler and R. Strom, Anal. Chem., 32:1890 (1960).
139.   M. P. Stevens and D. F. Percival, Anal. Chem., 36:1023 (1964).
140.   C. E. Bennett, A. J. Martin, and F. W. Martinez, Direct Analysis of Fatty
        Acids by Gas Chromatography, Presented at the 138th Meeting of the ACS,
        New York, 1960.
141.   J. J. Kirkland, Anal. Chem., 35:2003 (1963).
142.   E. M. Bens, Anal. Chem., 33:178 (1961).
143.   A. B. Littlewood, in collection: Gas Chromagoaphy 1958, D. H. Desty (ed.),
        New York, Academic Press, 1958, p. 23.
144.   C. Hishta, J. P. Messerly, and R. F. Reschke, Anal. Chem., 32:1730 (1960).
145.   C. Hishta, J. P. Messerly, R. F. Reschke, D. H. Fredericks, and W. P. Cooke,
        Anal. Chem., 32:880 (1960).
146.   D. H. Desty and C. L. A. Horbourn, Anal. Chem., 31:1965 (1959).
147.   A. W. Decord and G. V. Dineen, Anal. Chem., 32:164 (1960).

148. P. J. Pocaro and V. D. Johnston, Anal. Chem., 33:361 (1960).
149. F. R. Cropper and A. Heywood, Nature, 174:1063 (1954).
150. C. F. Cullis, F. R. F. Nardy, and D. W. Turner, Proc. Roy. Soc., A251:265 (1959).
151. Y. R. Naves, J. Soc. Cosmetic Chem., 9:101 (1958).
152. B. A. Rudenko and V. F. Kucherov, Dokl. Akad. Nauk SSSR, 145:577 (1962).
153. F. E. de Boer, Nature, 185:915 (1960).
154. J. Baum, J. Gas Chromatog., 1:13 (1963).
155. J. T. Kung and T. Romagnoli, Anal. Chem., 36:1161 (1964).
156. I. Sorensen and P. Soltoft, Acta Chem. Scand., 10:1673 (1956).
157. R. Ternishi and T. R. Mon, Anal. Chem., 36:1491 (1964).
158. V. Lukes, R. Komers, and V. Herout, J. Chromatog., 3:303 (1960).
159. S. Sunner, Mikrochim. Acta, 1956, p. 1144.
160. D. T. Sowyer and J. K. Borr, Anal. Chem., 34:1518 (1962).
161. A. I. Kestner, Zh. Fiz. Khim., 37:707 (1963).
162. S. E. Bresler, D. P. Dobychin, and A. G. Popov, Zh. Prikl. Khim., 36:66 (1963).
163 I. V. Borisenko, N. I. Bryzgalova, T. B. Gavrilova, and A. V. Kiselev, Neftekhimiya, 6:129 (1966).
164. V. M. Chertov and V. M. Belousov, Ukr. Khim. Zh., 31:171 (1965).
165. J. L. Webb, V. E. Smith, and H. W. Wells, J. Gas Chromatog., 5:387 (1965).
166. R. W. McKinney, J. Gas Chromatog., 5:389 (1965).
167. T. Brodasky, Anal. Chem., 36:1604 (1964).
168. G. F. Freeguard and R. Stock, Nature, 192:257 (1961).
169. A. J. B. Cruickshank and D. H. Everett, J. Chromatog., 11:989 (1963).
170. N. K. Bebris, A. V. Kiselev, and Yu. S. Nikitin, Kolloidn. Zh., 29:326 (1967).
171. N. K. Bebris, G. E. Zaitseva, A. V. Kiselev, Yu. S. Nikitin, and Ya. I. Yashin, Neftekhimiya, 8:481 (1968).
172. A. V. Kiselev, in collection: Proceedings of the Sixth Symposium on Gas Chromatography, Berlin, 1968 (in press).
173. Yu. S. Nikitin, Ibid.
174. E. Day, A. V. Kiselev, B. V. Kuznetsov, and Yu. S. Nikitin, Kinetika i Kataliz (in press).
175. J. F. Johnson and E. M. Barrall, J. Chromatog., 31:547 (1967).
176. A. A. Potatuev, A. I. Parimskii, and I. K. Shelomov, Zh. Anal. Khim., 22:463 (1965).

Chapter IV

Gas-Chromatographic Determination of Adsorption and Specific Surface for Solids

PHYSICOCHEMICAL APPLICATIONS
OF GAS CHROMATOGRAPHY

Gas chromatography finds its main use in analysis, but is also being widely used in reseach on processes in solutions and gases, and also at the surfaces of solids. Examples are determination of activity coefficients [1], entropies and heats of solution [2, 3], vapor pressure [4-6], molecular weights [7-9], diffusion coefficients [10, 11], adsorption isotherms [12-16],* and heats and entropies of adsorption [17-23]; also research on molecular interactions [24-26], measurement of specific surface [27-32], activation energies for internal diffusion [38], boiling points of high-boiling hydrocarbons [39], and gas–liquid interface resistance [40].

Gas chromatography in these applications differs from static methods in the rapidity of measurement with simple standard apparatus that does not need to be evacuated; moreover, the accuracy is often retained or even considerably increased. Further, the operations can be conducted at a wide range of temperatures. Gas chromatography differs from the calorimetric method in adsorption measurements in that very low θ and much higher T may be used [41]. In addition, adsorption isotherms can be measured for corrosive substances (in particular, ones containing fluorine [42], sulfur, etc.), which are difficult to record by static methods.

---

*See below for fuller details of the literature.

DERIVATION OF THE EQUILIBRIUM
ADSORPTION ISOTHERM
FOR AN EQUILIBRIUM CHROMATOGRAM

Wilson [43] first examined the quantitative relation of the equilibrium distribution between fixed and mobile phases to the equilibrium chromatogram. The work was extended by Weis and de Voult [44, 45] and (especially) by Gluckauf [46], who devised a method of deducing the equilibrium isotherm from the chromatogram, on the assumption that the conditions were those of equilibrium (instantaneous diffusion and instantaneous establishment of adsorption equilibrium). He demonstrated that this was so for liquid–solid chromatography [46].

Wicke [12], Cremer [13, 14], and James and Phillips [47] used this method to derive adsorption isotherms from gas chromatography. Gregg and Stock [15, 48] used this method to derive adsorption isotherms over a wide pressure range and for curves corresponding to different shapes of adsorption isotherm.

Gluckauf's method [46] is very simple and gives results rapidly, but it involves the assumption of instantaneous equilibration, so no allowance is made for broadening due to diffusion or to delay in transfer between phases. Vespalec and Grubner [49] used short columns to minimize diffusion broadening, with a finely powdered adsorbent. Schay [50-52] and Roginskii et al. [53] used Gluckauf's method with several recommendations to facilitate practical calculations.

In the theory of equilibrium chromatography it is assumed that conditions virtually rule out diffusion and kinetic forms of broadening, the sole cause of broadening being deviation from a Henry isotherm. This equilibrium (but not ideal) chromatography corresponds to the following equation for the material balance in an elementary layer [54, 55]:

$$-u_0 v \left(\frac{\partial c}{\partial x} \right)_t = v \left(\frac{\partial c}{\partial t} \right)_x + v_a \left(\frac{\partial c_a}{\partial t} \right)_x , \qquad \text{(IV, 1)}$$

in which u_0 is the linear velocity of the carrier gas, v is the volume of the gas phase in the column, v_a is the volume of the adsorbed layer, c is the concentration of adsorbate in the gas phase, c_a is the same in the adsorbed layer, t is the time elapsed from injection of the sample, and x is the distance from the start. This becomes [56-58]

$$u_0 v \Big/ \left(\frac{\partial x}{\partial t}\right)_c = v + v_a \left(\frac{\partial c_a}{\partial c}\right)_x \qquad \text{(IV, 2)}$$

in which $(\partial x/\partial t)_c = u_c$ is the linear velocity of a point of given c, while $v_a (\partial c_a/\partial c)_x$ in equilibrium conditions is the slope of the adsorption isotherm at the point corresponding to c, since here $(\partial c_a/\partial c)_x = (\partial c_a/\partial c)$ and $v_a dc = md a$ is the corresponding change in the uptake a (m is the mass of adsorbent). Then (IV, 2) gives

$$\frac{da}{dc} = \frac{vu_0 - vu_c}{mu_c} = \frac{v}{m}\left(\frac{u_0}{u_c} - 1\right) = \frac{v}{mt_0}(t_c - t_0) = \frac{w}{m}(t_c - t_0) = \frac{V_c}{m},$$

$$\text{(IV, 3)}$$

in which w is the flow rate of the carrier gas, t_c and t_0 are the retention times for the adsorbate (concentration c) and the carrier gas, and V_c is the retained volume of adsorbate at concentration c. Then the uptake with an equilibrium concentration c in the gas phase is

$$a = \frac{1}{m} \int_0^c V_c dc. \qquad \text{(IV, 4)}$$

Then (IV, 4) gives a as a function of c, i.e., the adsorption isotherm $a = \varphi(c)$, which is to be calculated via V_c and c expressed in terms of the quantities recorded by the chromatograph recorder.

It is necessary to express the detector reading in terms of c in order to derive the integral of (IV, 4) and also to deduce the c needed for $\dot{a} = \varphi(c)$ or $a = f(p)$, in which $p = cRT$ is the partial pressure of the adsorbate in the gas phase. The pen deviation h is usually proportional to c:

$$c = kh, \qquad \text{(IV, 5)}$$

in which k is a constant for a given compound and a given sensitivity range of the detector; it may be deduced by the frontal method from the relation of h to c. This method is reliable, but it requires the production of accurately known c in the gas; moreover, not all chromatographs are adapted for use with the frontal method. It is more convenient to use developmental chromatograms by injecting an accurately determined mass m_a [57]. On emergence

$$m_a = \int_0^\infty c \, dV, \qquad \text{(IV, 6)}$$

in which V is the volume of gas flowing through the column, while the limits of integration correspond to injection and complete elution. Then from (IV, 5) we have

$$m_a = k \int_0^\infty h \, dV. \qquad \text{(IV, 7)}$$

Usually the recorder is calibrated not in V but in length z along the chart. The chart speed is

$$q = \frac{dz}{dt}. \qquad \text{(IV, 8)}$$

The flow rate of the carrier gas is

$$w = \frac{dV}{dt}, \qquad \text{(IV, 9)}$$

so (IV, 8) and (IV, 9) give

$$dV = \frac{w}{q} \, dz. \qquad \text{(IV, 10)}$$

Substitution into (IV, 7) gives

$$m_a = k \frac{w}{q} \int_0^\infty h \, dz = k \frac{w}{q} S, \qquad \text{(IV, 11)}$$

in which the area of the peak is

$$S = \int_0^\infty h \, dz \qquad \text{(IV, 12)}$$

and thus is the area under the curve. The calibration constant then is

$$k = \frac{m_a q}{S_w}.$$
(IV, 13)

Substitution of $dc = kdh$ and $V_c = w(t_c - t_0) = (w/q)(z_h - z_0)$, into (IV, 4), in which $z_h - z_0$ is the distance on the chart from the point of emergence of an unadsorbed component to the point where gas of concentration c (deflection h) occurs; then

$$a = \frac{kw}{mq} \int_0^h (z_h - z_0)\, dh = \frac{kw}{mq} S_a,$$
(IV, 14)

in which

$$S_a = \int_0^h (z_h - z_0)\, dh$$
(IV, 15)

is the area on the chart between the axis of h at $z = z_0$ and the edge of the peak. Figure 60 illustrates determination of this area for peaks with broad trailing and leading edges, which correspond to adsorption isotherms free from inflection that are convex respectively to the uptake axis and to the concentration axis.

The w of (IV, 13) is the flow rate of the gas in the detector at the column temperature. Substitution of (IV, 13) for k into (IV, 14) gives

$$a = \frac{m_a S_a}{mS}.$$
(IV, 16)

This simple formula is usually employed to deduce the uptake from the chart. Here we must bear in mind the assumed constancy of w and q. There is also the dependence of w on the pressure difference across the column, as well as the corrections for the pressure and temperature changes between the column and the measuring head [58].

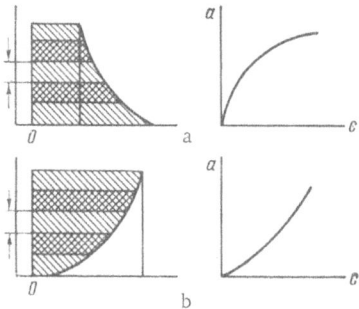

Fig. 60. Graphical integration and calculation of adsorption isotherms. Ordinate h, abscissa z: a) peak with sharp front and broad trailing edge; b) peak with broad front and sharp trailing edge.

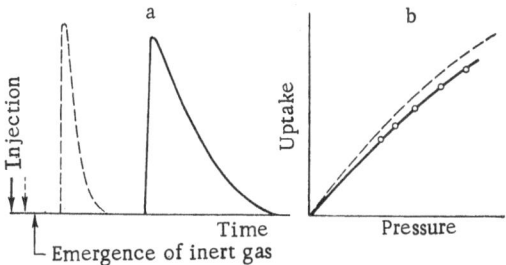

Fig. 61. a) Peaks of ethylene on activated charcoal at 101°C; b) adsorption isotherms. Full lines, column 1 m long containing 6.91 g of activated charcoal; broken lines, column 30 cm long containing 1.6 g of activated charcoal; O, data from static measurement.

It follows from (IV, 13) that the c corresponding to a is

$$c = \frac{m_a q h}{S w}$$ (IV, 17)

and the partial pressure is

$$p = \frac{m_a q h R T}{S w} .$$ (IV, 18)

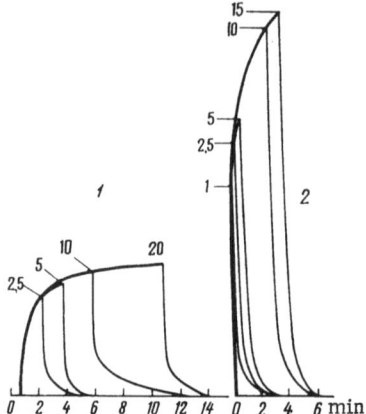

Fig. 62. Chromatograms of water on graphitized thermal carbon black for sample sizes
(μl) stated at: 1) 32°C; 2) 50°C; column
100 × 0.4 cm, helium 50 ml/min, katharometer, Tsvet-1 chromatograph.

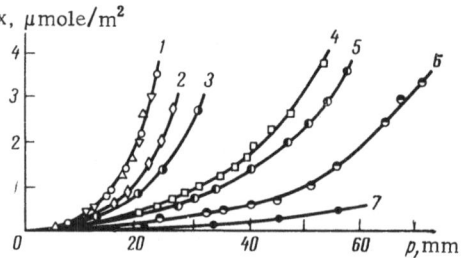

Fig. 63. Adsorption isotherms for water vapor on
graphitized carbon black deduced from the curves
of Fig. 62: 1) 29°C; 2) 32°C; 3) 34°C; 4) 40°C; 5)
43°C; 6) 50°C; 7) 60°C. The 29°C isotherm shows
points for different gas flow rates (40 to 60 ml/min).

This formula is used to determine the equilibrium pressure from
the recording. Hence, points on $a = f$(p) may be derived without
separate calibration of the detector if m_a is known exactly, while
w and q remain constant during the measurement, which yields h,
S, and S_a

In this way Huber and Keulemans [57] derived adsorption isotherms convex to the a axis for ethylene on charcoal at 100°C from

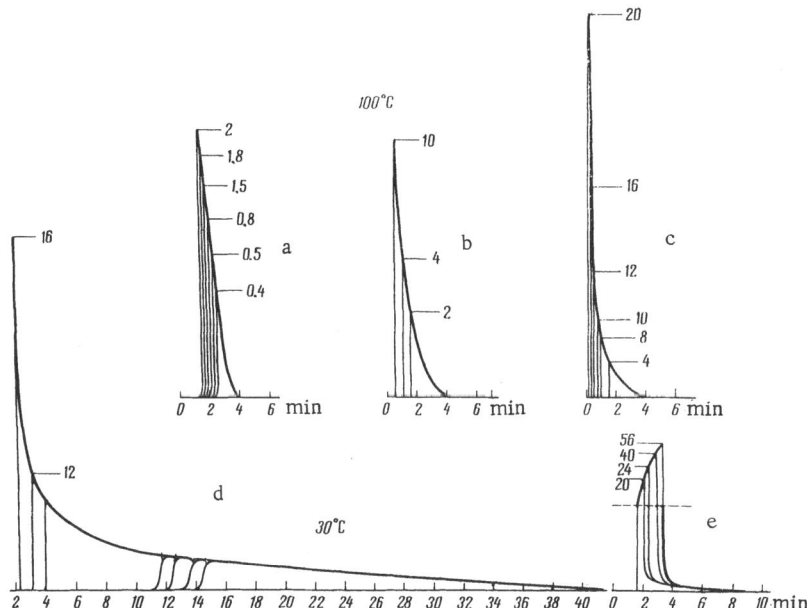

Fig. 64. Benzene on graphitized carbon black for various sample sizes and temperatures (conditions as in Fig. 62). The heavy lines are envelopes used in the integration.

chromatograms with trailing edges, the results being in satisfactory agreement with static measurements (Fig. 61).

Figure 62 illustrates the adsorption of water on the uniform
surface of graphitized carbon black [59] for different sample sizes
at two temperatures; Fig. 63 shows corresponding adsorption isotherms. Here the unspecific interaction with the solid is very weak,
while the water molecules interact strongly, so the isotherm is
convex to the pressure axis (Chapter II, p. 16), and the chromatogram has a broad leading edge. Broad trailing edges at small h
persist for the column without adsorbent, so here the integration
is over the broad leading edge. The points on each isotherm correspond to different h on the leading edge of the envelope of all the
peaks (shown by heavy lines in Fig. 62).

Derivation of isotherms with inflections is of some interest.
Gregg and Stock [15] first did this by the frontal method for the
cyclohexane-benzene system on silica gel, but the other method is
almost always more convenient. This will be considered for two

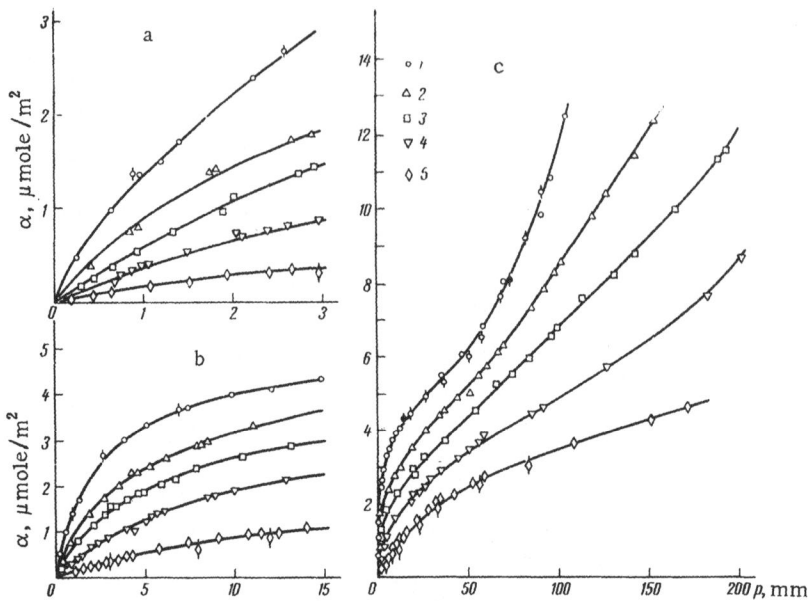

Fig. 65. Adsorption isotherms of benzene on carbon black at: 1) 30°C; 2) 40°C;
3) 50°C; 4) 70°C; 5) 100°C as derived from the curves of Fig. 64: a) for θ small;
b) medium θ; c) θ large. The points with lines through them represent static
measurements.

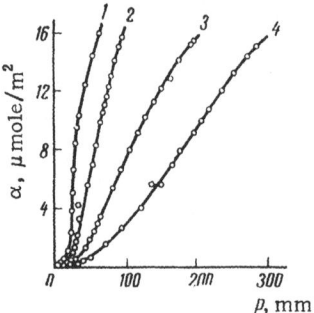

Fig. 66. Adsorption of methanol
on graphitized carbon black as deduced from the chromatograms of
Fig. 67 for: 1) 20°C; 2) 30°C; 3)
42°C; 4) 53°C. The points for 20°C
were determined by the static
method.

major types of isotherm with one
and two points of inflection. Figure 64 shows benzene on homogeneous graphitized carbon black
[59] (weak interaction between
molecules), where the isotherms
are of Brunauer's [60] second type
(Fig. 65) and are reasonably well
described by the Brunauer–
Emmett–Teller (BET) equation
[60, 61]. Figure 64 shows that the
shape of the curve varies with h,
the trailing edge being elongated
for h small (a-e in Fig. 64), which
corresponds to an isotherm convex to the uptake axis (when the
adsorption of benzene occurs

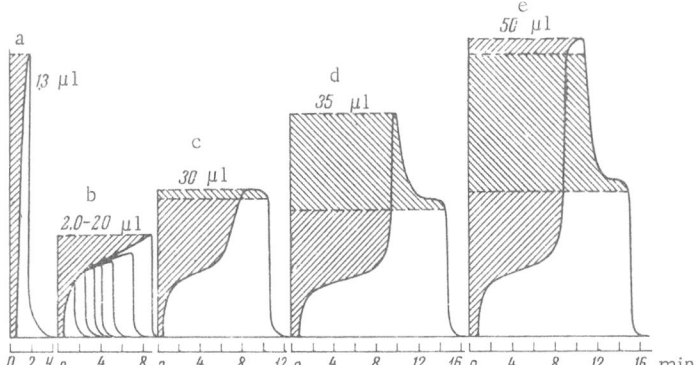

Fig. 67. Methanol on graphitized carbon black at 30°C for different
sample sizes (81 × 0.6 cm column, helium 50 ml/min, katharometer,
sensitivity reducing from a to e).

mainly in the first layer, Fig. 65). The leading edge is broadened
for h large (Fig. 64e), which corresponds to an isotherm convex
to the pressure axis past the inflection (uptake of benzene occurring mainly in the second and subsequent layers). The middle of
the isotherm (near the inflection) is almost linear, and on the chromatogram this is seen as a range where h increases with sample
size at a nearly constant z, i.e., both edges in this middle region
of h are nearly vertical (curves d). Figure 65 illustrates these
three regions.

The stronger adsorbate−adsorbate interaction for n-hexane
causes the adsorption isotherm for graphitized carbon black to be
at first convex to the pressure axis; there are subsequently two
points of inflection. This complicates the chromatograms [111].

As a criterion for equilibrium on the column one can use
either coincidence of the wings for different sample sizes or the
inclinations of the two sides of the peak. Diffusion is fairly rapid
in such cases and does not distort the peaks appreciably, the shape
then being determined by that of the equilibrium adsorption isotherm. Passage of an elongated wing from one side of the peak to
the other corresponds to a point of inflection on the isotherm; but
this effect is not very pronounced, which can give rise to error in
deriving the isotherm.

The next example concerns methanol on uniform graphitized carbon black [62], where the interaction with the solid remains unspecific, while being stronger than that for water (on account of the CH_3 group) but less than that for benzene. The methanol molecules interact strongly and specifically (association via hydrogen bonds, Chapter II, pp. 16-17), and so the isotherm is convex to the p axis at low p (Fig. 66). The specific interaction increases with α, and the isotherm rises more steeply; but there are not unlimited sites, so the uptake in the first layer largely ceases, which gives rise to the first point of inflection (Figs. 2 and 28). Uptake as second and subsequent layers occurs as p is raised further, and hence the same behavior is repeated, giving rise to a second point of inflection.

Figure 67 shows typical chromatograms for this case. At low T the isotherms (Fig. 65) have two parts convex to the pressure axis and a middle part concave to that axis; further, near $\theta = 0.5$ there is a thermodynamically unstable state (a single p corresponds to more than one α). Correspondingly, only the front edge is broadened at high T and the smaller h, while initially the front edge is broadened at low T and higher h, and then (near the region of discontinuity) a large part corresponding to several α for a given p, above which (at higher h) occurs a broadened trailing edge; finally at h above the second inflection we get a broadened front edge again.

The chromatograms of Figs. 61, 62, 64, and 67 show that proper choice of column length, gas speed, temperature, and relation of molecular size to pore size will prevent diffusion and adsorption exchange from appreciably distorting the peak, whose shape is mainly governed by the equilibrium isotherm. This is clear from the coincidence of the broadened edges for different sample sizes and the very nearly vertical sharp edges.

Only if conditions are favorable are the results from these gas-chromatographic measurements close to those from static adsorption [57, 59].

Gas chromatography allows one to measure the uptake over a wide range in T, which is very important for substance having low vapor pressures at room temperature and for research on adsorption by catalysts. Gas chromatography differs from static methods in coming closer to the actual conditions of adsorption on catalysts, i.e., adsorption of mixtures under dynamic conditions at

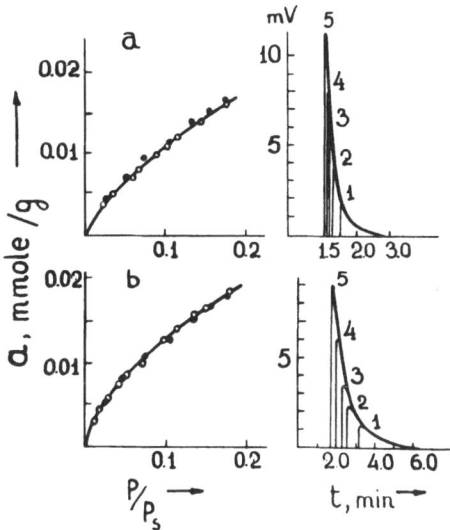

Fig. 68. Chromatograms and isotherms (open
circles) for benzene vapor at 25° on macroporous: a) silica gel; b) alumosilica gel. Sample
sizes in μmole: 1) 4.21; 2) 6.33; 3) 8.43; 4) 10.8;
5) 12.65. Filled symbols represent results from
static measurements.

high T. Cremer and Huber [41, 63] have recorded isotherms for
benzene and hexane on gels (silica, alumina, and aluminosilica) at
300–500°C in this way.

Gas chromatography gives good results for the adsorption
isotherms for macroporous gels. Figure 68 shows that the trailing edge is broadened for benzene (specifically adsorbed). The
isotherms derived from these chromatograms agree with those obtained by the static vacuum method [112]. Benzene has only weak
adsorbate–adsorbate interaction, so these isotherms are convex
to the uptake axis, as for adsorption on graphitized carbon black
(Fig. 65a and b). Macroporous silica gel adsorbs n-octane nonspecifically, and here the leading edge is broadened, as for n-
alkanes on graphitized carbon, because the isotherm for n-octane
on this extremely uniform surface is initially convex to the p axis,
on account of the substantial adsorbate—adsorbate attraction. The
trailing edges are almost vertical for large samples, but they have

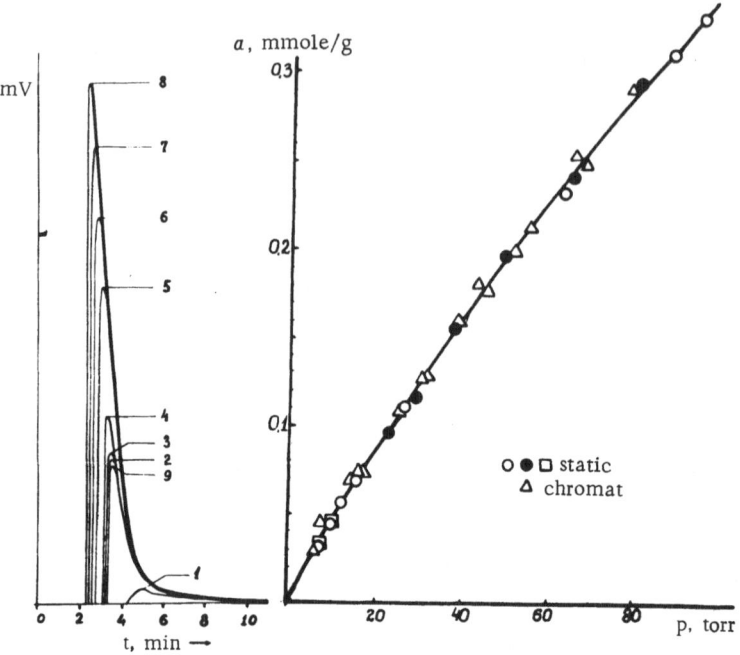

Fig. 69. Chromatograms and isotherms for CO_2 at 150° on NaX zeolite.

trails for low adsorbate concentrations in the carrier gas. The shape of the entire trailing edge is affected for small samples. This peak shape might be due to residual slight inhomogeneity, which causes the isotherm to be convex to the uptake axis for very small samples; but blank tests show that this broadening of the trailing edge occurs at low concentrations even when there is no silica gel in the column, and so the S_a of (IV, 16) should be calculated by integration over the leading edges, which coincide for samples of different sizes. Static measurements confirm this shape for the isotherms for n-alkanes on macroporous silica gel.

The same material, after partial dehydroxylation at 700° without change in s, gave increased V_s for n-alkanes at low θ (a flame-ionization detector was used). This is due mainly to reduction in the distance between the force centers in the molecules and the oxygen atoms in the silica. Group B compounds, which are specifically adsorbed at the OH groups, show reduced V_s after dehydroxylation. Firing of alumosilica gel at 700° increased V_s for all the compounds examined, except benzene, which is due to strong donor–acceptor interaction between the nonbonding pairs of the group B molecules and the aprotonic acid centers produced by the firing.

Gas chromatography is also of considerable interest in relation to research on adsorption by materials with small pores, in particular zeolites; it has been used with various inorganic gases and hydrocarbons for zeolites (see [113] for literature). The Q_0 for n-alkanes on NaX zeolite and on the outer surface of NaA zeolite crystals have been compared [114]. It was found that NaX gave some reduction in the gas-chromatographic Q for the n-alkanes as the chain length increased if these values were compared with those measured calorimetrically at 20°, although the reduction was not as large as that reported in [115] (see [114]). However, the gas-chromatographic Q for the larger n-alkanes were measured at substantially higher T [114], so the effect may be due not so much to slow diffusion of the large molecules within the small channels as to effects of T on the equilibrium Q. It is thus of interest to use gas chromatography in research on specific adsorption of CO_2 by zeolites, since CO_2 (which has a high quadrupole moment) is strongly adsorbed by zeolites [112, 116-118]. Figure 69 shows an example (due to Belyakova, Kiselev, and Keibal) for CO_2 at 150°C on crystals of NaX (without binder). The leading edges are almost vertical, while the trailing edges are broadened, though this does not vary with sample size, which indicates that the CO_2 comes virtually to equilibrium. However, the first peak in a series of equal samples of CO_2 on NaX was found to be smaller and to have a larger t_R than the later ones (e.g., samples 1-3 and 9, each 1 ml CO_2, in Fig. 69). NaX produces gradual irreversible adsorption of a little CO_2 (0.01-0.03 mmole/g at 150°C). NaA crystals fail also to release all of the CO_2 from the first batches (within the limit of detection); about 0.1 mmole/g is adsorbed irreversibly at 150°C. This effect is ascribed to chemisorption, which for CO_2 at low levels has been detected [119, 120] spectroscopically. This means that only completely reproducible peaks should be used in deriving the reversible isotherms for molecular adsorption of CO_2 via (IV, 16) and (IV, 18), i.e., a correction is in effect applied for the chemisorption of the CO_2.

Figure 69 shows the isotherm for reversible adsorption of CO_2 by NaX at 150°C as determined in this way; here again the results agree with those from static measurements, but below 100°C,

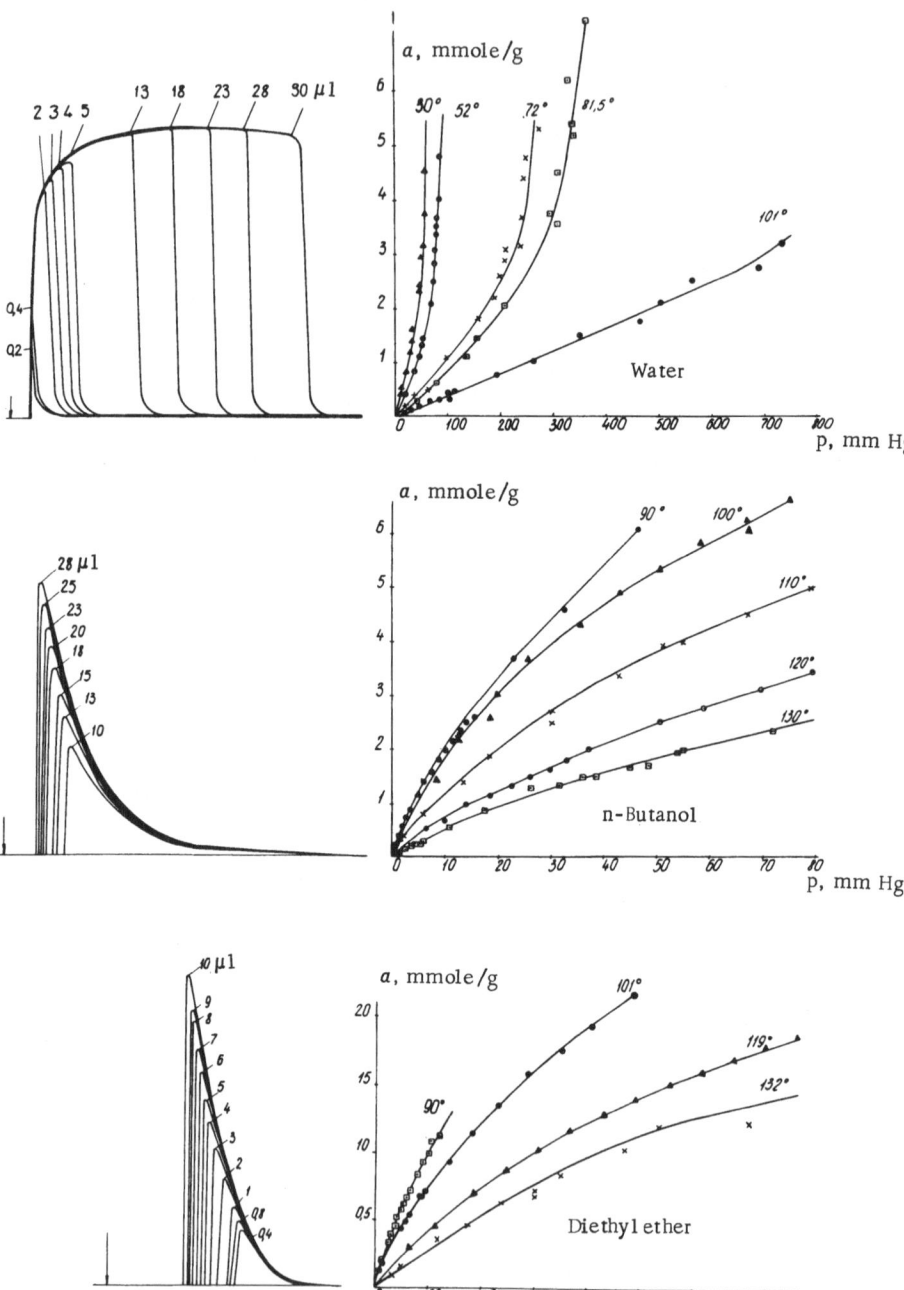

Fig. 70. Examples of chromatograms for various sample sizes and the resulting isotherms for Chromosorb 102 used at various temperatures with water, diethyl ether, and n-butanol.

where the uptake of CO_2 by zeolites is large, and exchange is slower, the leading edges are broadened, while the trailing edges do not coincide for different sample sizes. This shows that equilibrium is not attained. Formal calculation of isotherms from such curves then gives values lower than those from static measurements.

The same methods may be applied to porous polymers. It is difficult to measure the uptake of water by these polymers by static methods. Results for water in polyphenyldisiloxane show an extended leading edge, with the isotherm convex to the pressure axis. This, with the low uptake, indicates weak interaction with the polymer and strong adsorbate–adsorbate interaction. The Q for water and methanol on this material are below the heats of condensation, which is characteristic of adsorption on surfaces modified by chemical attachment of organosilicon groups [122], by deposition of close-packed layers [113, 123], or by a thick coating of solid organic material [124]. Figure 70 shows results for Chromosorb 102 [125], with the isotherms derived from different sample sizes. The edges coincide for different sample sizes, so conditions must be close to equilibrium. Isotherm measurement in such cases is greatly simplified by performing all calculations for the single peak from a large sample. The heavy lines in Fig. 70 link the curves defining the shape of the isotherm in this pattern of overlaping chromatograms for different sample sizes. Water gives a broad leading edge, as for graphitized carbon black, and the isotherm is concave to the pressure axis; whereas butanol and diethyl ether give broad trailing edges, and the isotherms thus are convex to the pressure axis, that for n-butanol (group D) rising more steeply than that for the ether (group B), since the OH groups interact much more strongly with this material, which is a weakly specific adsorbent of type III, on account of the phenyl groups.

Porous polyacrylonitrile has been examined in the same way [121]; here the functional groups have peripheral negative charge, and so group B compounds with high dipole moments (nitro-compounds and cyanides) are adsorbed fairly strongly. Group D compounds show more pronounced specific adsorption, as hydrogen bonds are formed to the CN groups. Here water gives a broadened trailing edge, which contrasts with the result for polyphenyldisiloxane and styrene—divinylbenzene copolymer (Fig. 70).

Gas chromatography has also been used to measure the uptake of reactive and corrosive substances [42], especially HCl and Cl_2 on glass [63].

Kipping and Winter [64] suggest that an adsorption isotherm should not be deduced from a single development curve, since sometimes the trailing edge varies with the sample size, as for water on silica. If this is so, gas chromatography cannot give good quantitative results, though it can give qualitative evidence. However, these difficulties are usually absent if the substance is not too strongly adsorbed, if care is taken in sample measurement and injection, and if the best gas rate is chosen.

Beebe et al.[65] have derived adsorption isotherms for nitrogen, argon, oxygen, CO, and hexafluoroethane on graphitized carbon by reference to frontal chromatograms. Schay [107] and Grubner [109] have used this method, which has the advantage of exact measurement of the adsorbate concentration in the gas; but the pulse method is simpler and more convenient for use with standard equipment.

GAS-CHROMATOGRAPHIC DETERMINATION
OF FREE ENERGIES AND OF HEATS
AND ENTROPIES OF ADSORPTION

Case of Narrow Symmetrical Peaks

Greene and Pust [17] first showed that heats of adsorption may be determined by gas chromatography if certain conditions are obeyed: if equilibrium is virtually attained and if the adsorption isotherm has an initial linear part (ideal equilibrium chromatography), then the slope represents Henry's constant:

$$\frac{dc_a}{dc} = K_{c_a,c} \,. \tag{IV, 19}$$

in which the subscripts to K indicate that the concentrations of the adsorbed and vapor phases are expressed in identical units. Then (IV, 3) implies that V_c (or V_R) is independent of c (narrow symmetrical peak) and

$$K_{a,c} = V_R/m = V_m, \tag{IV, 20}$$

in which V_m is expressed as cm^3 of vapor per g of adsorbent at the temperature of the column, while the subscripts to $K_{a,c}$ indicate that the uptake a is per g of adsorbent, with the content of the component in the gas phase referred to unit volume (e.g., mole per liter or mmole/ml).

To find K, and hence Q, from symmetrical peaks we need to determine the corrected retained volumes V_R from the maxima of narrow symmetrical peaks, the following formula [17, 58, 66] being used:

$$V_R = \frac{t_R w_{meas} p_{meas} T}{T_{meas} p_0} \cdot f = \frac{t_R w_{meas} p_{meas} T}{T_{meas} p_0} \cdot \frac{3}{2} \cdot \frac{(p_i/p_0)^2 - 1}{(p_i/p_0)^3 - 1}, \quad \text{(IV, 21)}$$

in which t_R is the corrected retention time, w_{meas} is the flow rate of the carrier gas in the flowmeter, p_{meas} and T_{meas} are the pressure and temperature in the flowmeter, and p_i and p_0 are the gas pressures at the inlet and outlet respectively.

The physicochemical constants independent of s (Ch. II, page 20) are [58]

$$K_{a, c} = K_{a, c}/s = V_s, \quad \text{(IV, 22)}$$

in which

$$V_s = V_m/s \quad \text{(IV, 23)}$$

is the retained volume per unit surface (nonporous and sufficiently macroporous).

If we have to convert from $\alpha = \varphi(c)$ to $\alpha = f(p)$, we have, since $p = cRT$, that

$$K_{a, p} = K_{a, c}/RT = V_s/RT. \quad \text{(IV, 24)}$$

From the equilibrium constant we get the standard change in the chemical potential μ (differential free energy) on adsorption as

$$\Delta\mu_0 = RT \ln K \quad \text{(IV, 25)}$$

It is more convenient to deduce $\Delta\mu_0$ from

$$\Delta\mu_0 = RT \ln (V_s/RT), \quad \text{(IV, 26)}$$

by using V_s in convenient absolute units (cm³/m²; see Table 2).

The isosteric heat of adsorption (reckoned positive if heat is released) is

$$Q_\alpha = RT^2 \left(\frac{\partial \ln p}{\partial T}\right)_\alpha = -R \left(\frac{\partial \ln p}{\partial 1/T}\right)_\alpha. \qquad (IV, 27)$$

This is independent of α in the Henry region:

$$Q = -RT^2 \frac{d \ln K_{\alpha, p}}{dT} = R \frac{d \ln K_{\alpha, p}}{d(1/T)} = -RT^2 \frac{d \ln V_s/T}{dT} = R \frac{d \ln (V_s/T)}{d(1/T)}. \qquad (IV, 28)$$

Now s is virtually independent of T, and m is quite independent of T, so Q may be deduced directly from the temperature dependence of V_m or V_R:

$$Q = R \frac{d \ln (V_m/T)}{d(1/T)} = R \frac{d \ln (V_R/T)}{d(1/T)}. \qquad (IV, 29)$$

Hence Q for narrow symmetrical peaks may be derived readily from V_s (or V_m or V_R) for several T (over the widest possible range in T) via a plot of log (V_s/T) [or log (V_m/T) or log (V_R/T) against $1/T$.

Figure 71 (upper part) shows curves for Xe on NaX zeolite [68] recorded at various T. These peaks are reasonably narrow because the inert gases interact nonspecifically with zeolites, while the atoms interact only weakly at small θ (Chapter II, Fig. 28 and p. 47). Figure 71 (bottom left) gives log (V_m/T) against $1/T$, while Fig. 71 (bottom right) gives log V_m against $1/T$ (V_m deduced from the maxima on the peaks of Fig. 66a). The bottom part of Fig. 71 shows that these relationships are linear (Q for small θ not strongly dependent on T). From (IV, 28) and (IV, 29) we have that the slope of the log (V_m/T) plot (etc.) against $1/T$ (bottom left in Fig. 71) multiplies by $2.303R = 4.575$ gives Q in cal/mole. To deduce Q from log V_s (or log V_m or log V_R) against $1/T$ (Fig. 71, bottom right) we multiply the slope by 4.575 and add $RT = 1.987T$ cal/mole (for the middle of the temperature range).

The NaX of Fig. 71 had the composition $Na_2O \cdot Al_2O_3 \cdot 3SiO_2$, and the result for Xe is 4.4 kcal/mole, while the Q for $\theta = 0$ of (II, 1)

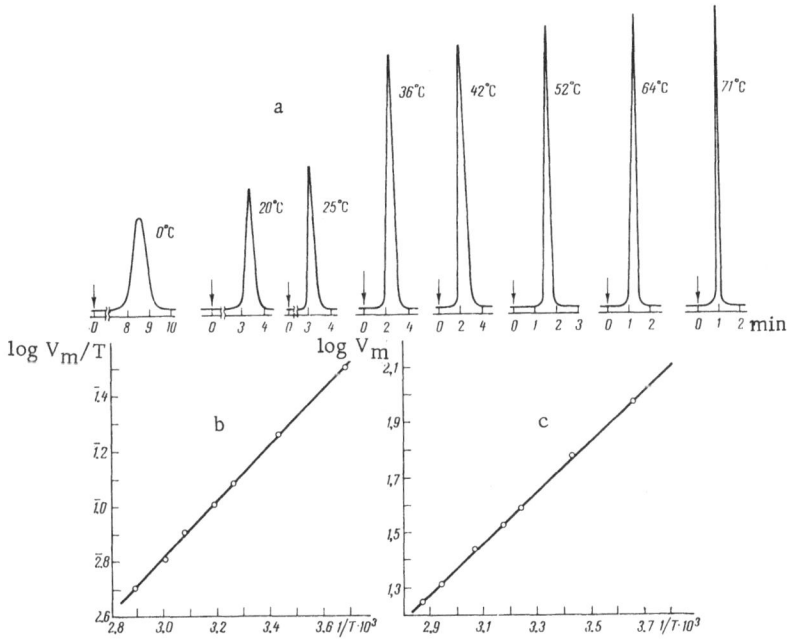

Fig. 71. a) Xe on NaX zeolite pressed into granules with binder (40 × 0.4 cm column, helium, katharometer, Tsvet-1 chromatograph) at several temperatures; b) $\log (V_m/T)$ against $1/T$; c) $\log V_m$ against $1/T$.

is 4.5 kcal/mole when the constant from vacuum static measurements are used (Chapter II, Fig. 28 and p. 47).

The entropy of adsorption (change in the differential entropy of the compound on adsorption on a given adsorbent) is

$$\Delta S_0 = \frac{1}{T}\,(\Delta U_0 - \Delta \mu_0) = -\frac{1}{T}\,(Q_0 + \Delta \mu_0), \qquad (IV, 30)$$

in which ΔU_0 is the corresponding differential change in the internal energy of the adsorption system and Q_0 is the differential heat of adsorption at $\theta = 0$. Substitution of (IV, 28) and (IV, 26) into (IV, 30) gives

$$\Delta S_0 = R\left(T\,\frac{d \ln V_s/T}{dT} - \ln (V_s/T)\right), \qquad (IV, 31)$$

with V_s in cm^3/m^2; also V_m or V_R may be used in the derivative in (IV, 31).

Figure 8 gives log V_S and ΔS_0 as functions of Q_0 at 150°C for various organic compounds adsorbed on graphitized carbon black. Straight lines have been drawn through the points for log V_S and ΔS_0 for the normal hydrocarbons. The points for all other substances lie around these lines, with no obvious dependence on the polarity or on the character of the functional groups [69, 70]. The approximate equations for these straight lines for various column temperatures are [70]

$$\log V_s (100°\,C) = -3.6 + 0.38 Q_0 = -3.2 + 0.61 n, \qquad (IV, 32a)$$

$$\log V_s (150°\,C) = -3.5 + 0.31 Q_0 = -3.2 + 0.50 n, \qquad (IV, 32b)$$

$$\log V_s (200°\,C) = -3.5 + 0.26 Q_0 = -3.2 + 0.41 n, \qquad (IV, 32c)$$

$$\Delta S_0 (150°\,C) = -16._0 - 0.92\, Q_0 = -16._9 - 1.47 n \qquad (IV, 33)$$

V_S is in ml/m², with Q_0 in kcal/mole, n being the number of carbon atoms in the molecule of the normal alkane. Equations (IV, 32) allow us to use the V_S to identify components from chromatograms on graphitized carbon. (In Fig. 8, ΔS_0 was calculated assuming $\Delta M_0 = RT \ln V_S$.)

Linear relations of ΔS_0 to Q_0 have been reported in [69–72]; they are characteristic of nonspecific adsorption, at least. Theoretical numerical estimates have been given [73] for the relation of ΔS_0 to Q_0. It has also been shown [74] that the polarity of the solute has little effect on the entropy of solution in a nonpolar solvent.

Only narrow peaks (with both edges in the ideal case vertical, and the width vanishingly small) correspond to ideal equilibrium chromatography on a uniform surface in the absence of adsorbate – adsorbate interaction. Hence a single symmetrical peak is, generally speaking, inadequate for calculation from ideal equilibrium thermodynamic formulas.

In particular, this can be seen for the adsorption of normal alkanes on the nonporous uniform surfaces of graphitized carbon black and of the channels in NaX zeolite [75]. In both cases the peaks are symmetrical, but for NaX the peaks rapidly broaden as n increases, because of retarded exchange of the large molecules. These symmetrical but very much broadened peaks correspond to processes deviating so far from equilibrium that it is impossible to apply equilibrium formulas. Hence we would expect thermodynamic parameters calculated from such peaks via the equilibrium theory to differ from those measured under static conditions, and that the deviations would increase with Q_0 and with the molecular size. Figure 31 shows the heat of adsorption for the normal alkanes, as determined in the two ways, as a function of n for graphitized thermal carbon black and NaX. In the first case, the deviations are small and arise solely because gas-chromatographic experiments for n large are run at higher temperatures than static ones (see p. 130 for details); whereas the results for the porous crystals are in agreement only for low n, with low Q_0 from gas chromatography for the higher n.

Case of Unsymmetrical Peaks

Here the shape is dependent on that of the isotherm (pp. 109-120), so it is incorrect to deduce Q from the position of the peak as a function of temperature, incorrect even for a fixed sample size. Carberry [76] has examined this question; he showed from the balance equations that derivation of Q directly from the maxima via (IV, 29) is incorrect when the isotherms are curved, in particular for Freundlich isotherms.

Hence the isotherms (Fig. 62-67) should be deduced by the method given in the second section of this chapter (pp. 105-109) and from these are constructed the plots of log p against 1/T with a as

parameter, which give $\quad Q_a = -R\left[\dfrac{\partial \ln p}{\partial (1/T)}\right]_a$.

Figures 73-75 show Q_a as a function of surface concentration α derived from Figs. 63, 65, and 66, for water, benzene, and methanol on graphitized thermal carbon black (the curves may also be shown as functions of $\theta = \alpha/\alpha_m = \alpha\omega_m$, in which α_m is the capacity of a close-packed monolayer and $\omega_m = 1/\alpha_m$ is the area per molecule in such a layer). Figure 74 shows that Q_a for benzene as a function of α as deduced from chromatograms (e.g., Fig. 64) is close to that derived from calorimetric measures. Analogous results have been obtained for n-hexane on graphitized carbon black, where adsorbate—adsorbate interaction causes Q to increase appreciably with α; Fig. 72 shows this relationship as deduced from chromatograms, which is close to that obtained calorimetrically at 20°C [126] and almost coincides with the $Q = f(\alpha)$ curve recalculated from 20°C for a mean column temperature of 70° via the heat capacity measured calorimet-

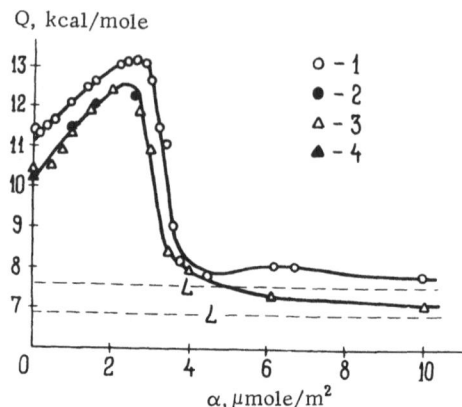

Fig. 72. Differential heat of adsorption for n-hexane on graphitized carbon black as a function of
surface filling from: 1) measurement in a calorimeter at 20°C; 2) chromatographic determination;
3) conversion of points 1 from 20 to 70° via the thermal capacity; 4) calculation from the dependence
on K_1 on T.

rically [127]. It is possible to use

$$Q = Rd \ln V_m/d(1/T)$$

to determine Q (with V_m the retained volume for the peak maximum)
only if the peaks are narrow and symmetrical, i.e., very small
samples and adequately high T. If the peaks are unsymmetrical,
it is first necessary to derive the adsorption isotherms for several temperatures, which are then used to determine Q as a function of α.

This method thus allows one to examine the equilibrium adsorption when the adsorbent is nonporous and nonspecific. The Q
for water are small, so α for graphitized thermal carbon black remains
very small (the α correspond to filling of only a small part of the
surface even for $p \rightarrow p_s$). The hydrocarbon radical in an alcohol
facilitates attachment to graphitized carbon, and the Q increase
rapidly, on account of adsorbate–adsorbate interaction via hydrogen bonds.

Gas chromatography allows the very low Q_0 for water and
methanol to be measured; here there is no adsorbate–adsorbate

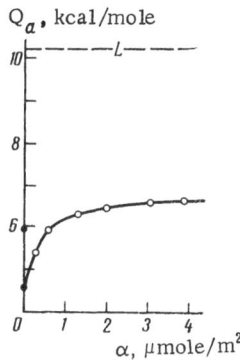

Fig. 73. Relation of Q_a to α from gas chromatograms of Figs. 62 and 63 for water on graphitized carbon black [59]: •) chromatographic data of [77]; L) the heat of condensation.

interaction, though this makes a large contribution at higher θ. Hence, gas chromatography allows us to deduce the energies of hydrogen bonds in monolayers on nonspecific adsorbents (see pp. 134-136).

It is of interest to compare the results from static methods with those derived from gas chromatography on the assumption of equilibrium, as in Fig. 34 and Table 2; the results are closely similar for homogeneous surfaces, especially for nonporous and nonspecific graphitized thermal carbon black [59, 62]. Good agreement is also obtained for the relatively weakly adsorbed inert gases and lower hydrocarbons on the geometrically very uniform crystals of zeolites (Fig. 31).

Satisfactory agreement has also been obtained for inhomogeneous surfaces (e.g., Fig. 61), but only for relatively weakly adsorbed substances [57], the results for strongly adsorbed compounds usually being worse. For example, the Q found [79] for the $C_5 - C_8$ normal alkanes and benzene on silica gels with pores of 100 to 1000 Å are 15-20% below the calorimetric values for the same θ.

There are the following possible reasons for these discrepancies: 1) deviation from equilibrium; 2) effects of compounds previously taken up from the air during injection of sample (water, ammonia, CO_2); 3) impurities adsorbed from the carrier gas (e.g., water) on account of difficulty of removal (see p. 244); 4) adsorption of carrier gas on the most active areas; 5) use (commonly) of higher temperatures in gas chromatography than, in particular, in calorimetric measurements; 6) effects of adsorbate−adsorbate interaction, which is more pronounced in static measurements at high θ and low T.

Causes 2) and 3) (small polar molecules) are unimportant with nonspecific adsorbents [59, 62, 75]; Fig. 31 shows that static and gas-chromatographic methods give similar values for the normal alkanes on graphitized carbon black. The effect should also not be

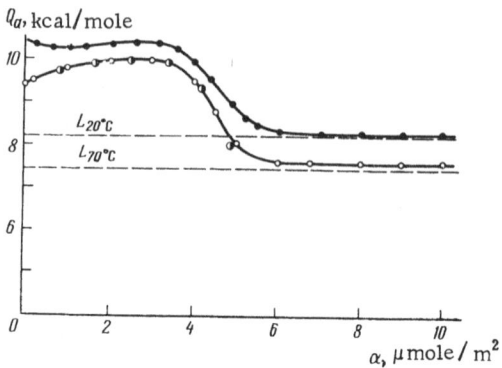

Fig. 74. Relation of Q_a to α for benzene on graphitized carbon black: O) from chromatograms of Fig. 64 and isotherms of Fig. 65; ●) calorimetric data for 20°C [78]; O) Q_a at 70°C deduced from calorimetric data via the heat capacity of adsorbed benzene (see Fig. 77).

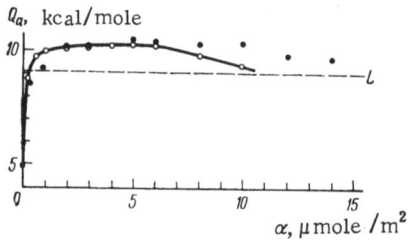

Fig. 75. Relation of Q_a to α for methanol on graphitized carbon black: ●) from chromatograms of Fig. 67 and isotherms of Fig. 66; O) from static measurements.

important with specific adsorbents with homogeneous surfaces and large s, since only a small fraction of the surface can be poisoned in the time needed by a chromatography run. Conversely, a strongly specific adsorbent of low s has both causes very important, as for MgO [80] and TiO_2 [81] of low s. Hence, difficulties occur in gas-chromatographic study of the uptake by the clean surface in such cases, at least for low T. Even a specific adsorbent of high s but lacking homogeneity is susceptible to poisoning of the most active sites.

Cause 4) may be eliminated by using hydrogen or helium carrier. Cause 6) is important only for group D compounds on nonspecific adsorbents at low T (i.e., more under static conditions), the effects being slight for group D on specific adsorbents. Hence, these factors may be eliminated, or at least largely so, by using nonspecific adsorbents at fairly high T, or if the specific ad-

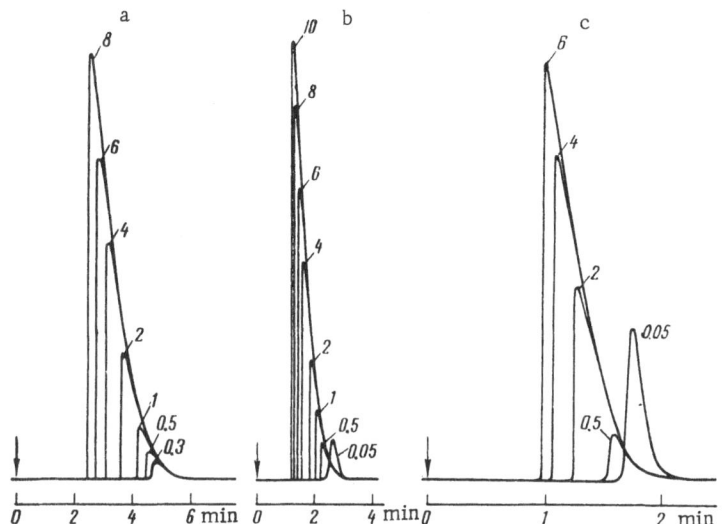

Fig. 76. Propane on silica gel at: a) 30℃; b) 50℃; c) 70℃ (column 100 × 0.5 cm, 0.25-0.5 mm grains, helium 50 ml/min, katharometer). The sample sizes (ml) are given at the peaks. A flame-ionization detector was used for the 0.05 ml sample.

sorbent is selected to be uniform and of high s, and also is used at fairly high T.

Cause 1) is eliminated by the use of nonporous or macroporous (or surface-porous) adsorbents; but then s becomes small, which for specific adsorbents may mean substantial blocking by water, etc., derived from the air (during filling) or the carrier gas. High T with small s may result in unduly small t_R, which results in poor accuracy for V_R and Q. This approach is therefore entirely reliable only for nonspecific adsorbents.

The effects of inhomogeneity and T on the uptake and on Q must be considered together. Figure 76 illustrates this for propane on silica gel [82], whose inhomogeneity causes the isotherms to be concave to the pressure axis (Figs. 39 and 60a), so the chromatograms have broadened trailing edges. The leading edges are practically vertical, while the trailing edges coincide. Curvature of the equilibrium isotherm is the cause of broadening in this case.

The isotherms derived by the above method (pp. 105-109) then allow one to deduce Q_a as a function of α. The isosteres from static

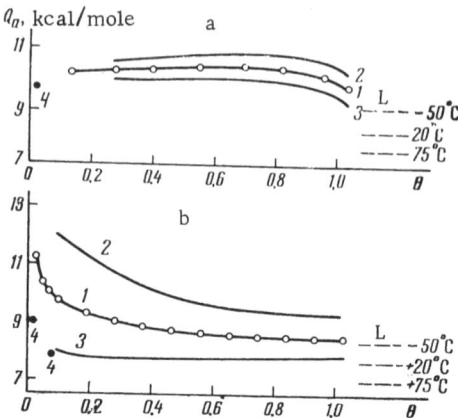

Fig. 77. Relation of Q_a to θ at several T for different surfaces: a) benzene on graphitized carbon black; b) n-hexane on silica gel; 1) calorimetric at 20°C; 2) and 3) calculated from 1) via measured heat capacities; 4) gas-chromatography data for 50-200°C (results of Fig. 74 in the case of benzene).

measurements usually relate to the lower T, while those from chromatograms relate to higher T, which raises the questions of the dependence of Q on T and of the effects of the inhomogeneity on this dependence. Measurements of the heat capacity facilitate answers to these questions as for benzene on graphitized carbon black [83] or n-hexane on silica gel with s = 340 m²/g and d≈100 Å in the pores [84]. In both cases the adsorbed compound has a heat capacity higher than that for the vapor state, so Q has a negative temperature coefficient. In both cases the heat capacity approaches that of the liquid as θ increases, but the precise trend is dependent on the adsorbent, the thermal capacity of benzene adsorbed on homogeneous graphitized carbon black being less than that of the liquid but increasing with θ, while n-hexane on inhomogeneous silica gel has a heat capacity above that of the liquid, which decreases as θ increases. Hence, Q for benzene for θ in the monolayer region falls less rapidly as T increases than does the heat of condensation, whereas Q for n-hexane falls more rapidly. In general, Q falls especially rapidly at small θ for inhomogeneous surfaces, as the effects of inhomogeneity are pronounced at low T.

Gas chromatography therefore gives Q only slightly lower than those from calorimetry in the case of strong nonspecific adsorption on uniform surfaces, but the results are very much lower when the surface is very inhomogeneous.

The main cause of the differences (except for specific adsorbents of low s) is that the two types of measurement are made at different T, and the effects of this are the more pronounced the greater the inhomogeneity. Figure 77 shows results for benzene on graphitized carbon black [83] and n-hexane on silica gel [84], which indicate that, at reasonably high T, gas chromatography gives Q_a close to those from static measurements reduced to the same T. Here the inevitable slight deviation from equilibrium has little effect.

However, a marked increase in T causes a fall in the equilibrium constants (V_R for θ small) and thus a loss of selectivity, so there is some optimal T. Alternatively, a more homogeneous adsorbent may be used, or the column temperature may be varied during a run.

The following are some examples of gas-chromatographic determination of Q. Greene and Pust [17] first used the method with Ar, O_2, N_2, CO, and CH_4 on activated charcoal; the results were compared with those from calorimetry. Q was also measured for light hydrocarbons and CO_2 on silica gel and alumina gel, and for Ar, O_2, N_2, Kr, and CH_4 on 5A zeolite. It was found that CO_2 was adsorbed irreversibly by alumina gel. The gas-chromatographic results were in satisfactory agreement with calorimetric ones.

Habgood and Hanlan [19] examined the properties of activated charcoal in this way, measuring V_R at various T and deducing Q, and finding that the last varied with θ and also somewhat with T, the Q for small samples being almost double those for large ones. The reason is that a small sample is adsorbed only on the most active sites (small pores).

The nature of the surface of an adsorbent can be rapidly established by measuring Q for a range of compounds.

Petrova et al. [20] determined Q for the C_1 to C_4 hydrocarbons on CaA zeolite; comparison with calorimetric values and with ones deduced from isosteres for static measurements shows that gas chromatography is a suitable method for Q if the adsorption on the zeolite is not strong. It is also suitable for examining the effects of molecular and surface structure. The Q for normal alkanes and alkenes on zeolites increase in proportion to the number of C atoms, and also with the appearance of unsaturation, on account of the more specific (than dispersion) interaction of the π-bonds with the cations in the channels.

Kiselev et al. [22] confirmed this by gas-chromatographic determination of Q for O_2, N_2, CO, the C_1-C_3 hydrocarbons, benzene, and n-hexane on zeolites of types CaA, CaX, and NaX. The Q for nitrogen, ethylene, and benzene substantially exceeded the corresponding values for oxygen, ethane, and n-hexane, on account of the specific interaction of the π-bonds with the cations. Also, the gas-chromatographic Q (based on the assumption of equilibrium) began to deviate from the static ones [75] as the size and adsorption energy of the hydrocarbons increased (Fig. 31). The high adsorption energy and directional specific interaction of nitrogen and ethylene cause the t_R (relative to methane and propane) to decrease as T increases. The specificity of the adsorption is much reduced by moistening the zeolite [22]. The ratio of the corrected t_R for ethylene to that for propane falls rapidly as the water constant increases, whereas the ratio of these times for ethane and propane is almost unaffected.

The effects of water content of CaA (5A) zeolite have also been examined as regards V_R and Q for H_2, O_2, N_2, CH_4, CO, Kr, and Xe; in every case t_R was inversely related to the water content, but the effect was more marked for H_2, O_2, N_2, CH_4, and CO than for Kr and Xe. In particular, the Kr and Xe peaks may tend to overlap adjacent ones as the water content varies. The gas-chromatographic Q for Kr and Xe increase with the amount of previously adsorbed water, while those for N_2 and CO decrease and those for O_2 and CH_4 remain nearly unchanged.

Habgood [86] measured V_R and Q by gas chromatography to examine the behavior of various cationic forms of the X zeolites (with Li^+, Na^+, K^+, Mg^{++}, Ca^{++}, Ba^{++}, and Ag^+). The Q for NaX agreed with published values for O_2, N_2, CH_4, and C_2H_6 but were somewhat low for C_3H_8 and C_4H_{10}. The interaction energy is inversely related to the radius of the cation. The interaction was strongest for AgX, ethylene in particular forming an unstable (chemisorbed) complex.

Very low Q have been reported [23] for Ar, O_2, N_2 CO, CO_2, and C_1-C_4 hydrocarbons on CaA (5A), NaX (13X), silica gel, activated charcoal, and alumina gel.

The Q_0 for C_1-C_{10} hydrocarbons on silica gels varying in porosity have [21] been measured by gas chromatography; Q_0 was inversely related to mean pore radius and directly to number of carbon atoms n (in agreement with static measurements) for small pore sizes; the relation to n was linear for the more macroporous adsorbents. However, it has been shown [79] that the Q_0 for saturated hydrocarbons and benzene on silica gel are less than the calorimetric values (p. 130).

The Q for n-alkanes on macroporous silica gel and alumo-silica gel [112] increase nearly in proportion to the number of CH_2 groups (by about 0.8 kcal/mole for $\theta \approx 0.1$); the Q for alumosilica gel at this θ are about 1 kcal/mole higher than those for silica gel. The difference for benzene or toluene is about 1.5 kcal/mole, as against about 2, 3, and 4 kcal/mole respectively for ether, acetone, and acetonitrile (group B compounds with nonbonding electron pairs). The difference is due to the greater acidity (protonic and aprotonic) of the latter gel. The two gels give closely similar Q for nitro-benzene (the large dipole moment is localized on the internal bonds in the molecule) [112]. The Q for pyridine, aniline, and (especially) aliphatic amines on silica gel exceed those for benzene and the corresponding n-alkanes by factors of two or more, but the adsorption is reversible; whereas alumosilica gel adsorbs pyridine and other amines irreversibly [128, 129], and so Q cannot be determined by gas chromatography. However, the irreversibility has been used to modify the surface for the purposes of analysis [129].

Gas chromatography has been used [75] to give Q_0 for normal alkanes on various materials; in some cases (macroporous silica gel, graphitized thermal carbon black, and crystals of NaA and NaX zeolites) these heats are doubled by a factor 2 for a given hydrocarbon (Fig. 35). We have seen in Chapter II that variation of the surface provides wide control of the selectivity of columns for non-specifically adsorbed components, not only in isothermal working but also with temperature variation.

Q values for hydrocarbons on alumina have been determined in this way [87], and also Q_0 for isobutene on alumina gel [88]. Q for CO on poisoned nickel oxide has been measured [89].

Q for various compounds on graphitized carbon blacks have been determined by Ross et al. [90], Chirnside and Pope [91], Gale and Beebe [77], Belyakova et al. [25, 26, 59, 62, 70], Kiselev and Yashin [71], and Petov and Shcherbakova [92]. These results show that Q for graphitized carbon black is almost independent of the local electron-density distribution (π-bonds, nonbonding pairs, dipole and quadrupole moments) but is dependent on the molecular geometry (pp. 16-40 of Chapter II). Q as a function of θ for benzene, water, and methanol on carbon black has [59, 62] been deduced from complex peaks (see Figs. 61-67, 73-75, and 77, pp. 111-114 and 127-130).

The accuracy of gas-chromatographic Q is dependent on that of t_R, T, gas flow rate, gas pressure, etc. Many analytic chromatographs do not allow T to be measured accurately (many columns operate with a longitudinal temperature gradient), nor do they allow the gas pressure and flow rate to be measured. A description has been given [93] of a system especially designed for exact physicochemical studies, in which the inlet pressure can be measured to ±0.1 mm Hg and the flow rate to ±0.1 ml/min, and the temperature differences along the column are only ±1°C.

This shows that gas chromatography may be used to advantage in examining the properties of any adsorbent; its high sensitivity makes it very valuable for solids with s <0.1 m^2/g [130] or with weak adsorption. Moreover, it can be used with substances of low vapor pressure. Comparison with static measurements shows that the pulse method (used with Gluckauf's theory) gives reproducible results for the Q and the isotherms, provided that certain conditions are met (pure carrier gas, T adequately high for the particular system, absence of chemical reaction with the surface).

Gas chromatography will be extended as a method of research on adsorption, which will stimulate the relevant theories, if the above characteristics are accumulated for a variety of compounds and adsorbents over a wide range in temperatures, especially with proper selection of geometry, electronic structure, chemical nature of surface, etc.

GAS-CHROMATOGRAPHIC DETERMINATION OF THE ENERGIES OF HYDROGEN BONDS BETWEEN ADSORBATE MOLECULES

Sensitive detectors (especially ionization ones) allow small θ to be used, and then graphitized thermal carbon black does not show adsorbate–adsorbate interactions, even when these are strong specific ones involving hydrogen bonds, as Belyakova et al. [25, 26] have shown for alcohols. There was a slight tendency for the trailing edge to broaden as the number of C atoms increased, but the position of the peak did not vary with the (small) size of the sample, so the isotherms were close to Henry ones at these small θ and fairly high T. Figure 78 shows Q_a as a function of n for normal hydrocarbons and alcohols as determined statically at small θ (in a calorimeter and otherwise, see [25] for literature) and as determined gas-chromatographically at very small θ and higher T. The figure also shows the calculated potential energies of adsorption for $\theta = 0$, namely Φ_0 (see [25] for literature). The Q_a increase nearly linearly with θ in monolayers of normal hydrocarbons, so Q_0 is readily determined by extrapolation to $\theta = 0$. Alcohols show a more complicated relation of Q_a to θ (Fig. 75), so it is difficult to deduce Q_0 from calorimetric data for the higher θ. Hence the Q_a of Fig. 77 for alcohols are given for $\theta \approx 0.1$; the Q_a for $\theta \approx 0.1$ for the normal alkanes are extremely close to Q_0 as found by extrapolation to $\theta = 0$.

Figure 78 shows that normal alkanes, whose molecules interact only weakly (and nonspecifically), give similar Q_0 by both methods, and these are also close to the theoretical $-\Phi_0$ for isolated molecules (denoted by crosses in Fig. 78). The normal alcohols (compounds of group D) show specific interaction, and only gas chromatography gives Q_0 close to $-\Phi_0$ as calculated for isolated molecules. The calorimetric values Q_a [94-96] relate to much higher and lower T (20°C); they are much larger because they include a contribution from the specific interaction between molecules, which involves hydrogen bonding. The difference between the two sets of results in Fig. 78 is 4.8 kcal/mole, which is the energy of the hydrogen bond between adsorbed alcohol molecules.

This example shows that gas chromatography and calorimetry may be combined for group D compounds to determine the hydrogen-bond energy for adsorbed layers on nonspecific adsorbents. Further, symmetrical peaks can be obtained even from strongly associated compounds by using type I (nonspecific) adsorbents (in particular, graphitized carbon black) in gas-chromatography columns with sensitive detectors and high T, because the low θ minimize interactions between molecules.

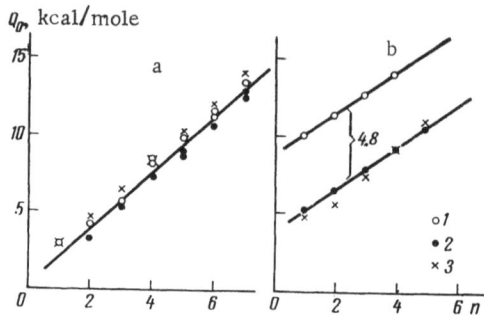

Fig. 78. Relation of Q_0 to n for graphitized carbon black for: a) normal alkanes; b) normal alcohols; 1) calorimetric results and values from isosteres derived by a vacuum static method; 2) gas-chromatography values; 3) theoretical values for potential energy of interaction of isolated molecules with an infinite graphite lattice.

Larger samples and lower T allow association effects to be detected in gas chromatography on graphitized carbon black. Figure 67 [62] shows methanol in this case for various sample sizes; the peaks are of complicated form because the interaction is strong and polymolecular adsorption predominates (Fig. 66).

GAS-CHROMATOGRAPHIC DETERMINATION
OF SPECIFIC SURFACE

Here there are the following two groups of methods:

1. Calculation from V_S as measured for adsorbents of the same nature and with known s.

2. Derivation of isotherms from chromatograms and deduction of s by the BET method [60, 61].

In the first case, s is deduced from (IV, 23), which gives

$$s = V_m/V_s,$$

(IV,34)

in which V_m is determined by chromatography, while V_S is already known.

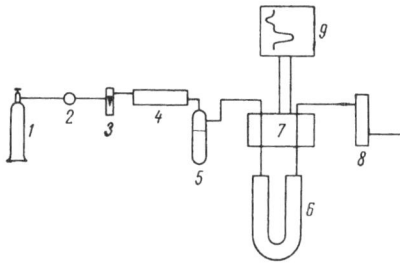

Fig. 79. Apparatus for measuring s by thermal
desorption: 1) cylinder; 2) valve; 3) rotameter;
4) dryer; 5) zeolite trap; 6) adsorbent tube; 7)
detector; 8) flowmeter; 9) recorder.

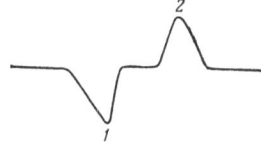

Fig. 80. Peaks of: 1) adsorption; 2) desorption of nitrogen
in the thermal desorption method.

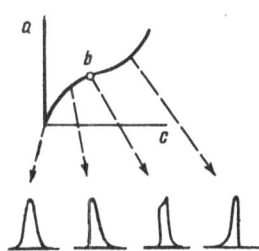

Fig. 81. Adsorption isotherm
(Brunauer's [60] type II) and the
corresponding shapes of peak
(only the upper part of the chromatogram is shown in the case of
the last peak; see Fig. 64).

This formula can be used only if the geometrical or chemical structure of the surface relating to V_m is similar to that relating to V_s (i.e., only for nonporous or macroporous adsorbents of identical chemical nature) and if the θ are similar (when V_s is strongly dependent on θ) [29]. Figures 3 and 41 show examples of the dependence of V_m on s for various adsorbents; V_m is proportional to s, and the slope is V_s.

The first demonstrations of this method are to be found in [27-30].

Rapid deduction of s requires measurement of V_m for a sufficiently strongly adsorbed vapor (the adsorption should be nonspecific, to minimize the effects of chemical inhomogeneity in the surface) on a column containing a mass m of adsorbent, V_s being taken from tables or previously determined for the same T for an adsorbent of the same nature at a similar θ. Here m, the sample size, and T must be chosen so that t_R can be measured precisely, while the peaks are narrow and symmetrical [29].

The BET method uses unsymmetrical peaks and requires derivation of equilibrium isotherms (see pp. 105-11); an example is [30], where gas-chromatographic isotherms were used with the BET method of [60, 61].

A third method, widely used, is called thermal desorption; it was developed in 1958 by Nelson and Eggertson [97] and Grubner [98], and was afterwards used by various workers [99-104]. The method is very simple and sensitive; s from 0.01 to 600 m^2/g can be measured [104].

Figure 79 shows the system for this method. The fine-control valve 2 is used to set a specified flow rate for the $He-N_2$ mixture. The gas passes through the filters 4 and 5, the reference arm of the katharometer 7, the column 6, the sample arm of the katharometer, and the flowmeter 8.

The surface is brought into equilibrium with the gas at room temperature, and then the column is placed in liquid nitrogen. Nitrogen is adsorbed, which alters the composition, which is recorded by the detector. This alteration ceases when adsorption is complete, and the recorder pen returns to the zero line. Then the Dewar is taken away and the column is rapidly heated to room temperature; the nitrogen is released, and a reverse peak is produced (Fig. 80). The uptake at the low temperature is usually deduced from the second peak, since this is more symmetrical (the desorption is very rapid during the heating).

Mixtures with 5, 10, 15, and 20% nitrogen in helium have been used to measure these isotherms in the region of transition to polymolecular adsorption.

First the apparatus is calibrated (the peak is measured as a function of N_2 concentration). Then the uptake is measured at a series of partial pressures; the adsorption isotherm is derived, and s is deduced by the BET method.

Sometimes there is no need to record the entire isotherm. If the absolute uptake (per unit surface) is known, i.e., the surface concentration α in the monolayer, the calculation is simplified, since α for the given nitrogen concentration in the desorption run can be taken from the isotherm. The ratio of the result a from thermal desorption to α (p. 121) for the same nitrogen pressure gives s [105].

However, it must be borne in mind that the area ω_m taken up by a nitrogen molecule in a monolayer is rather strongly dependent on the chemical nature of the surface [106], as is $\alpha_m = 1/\omega_m$ (the monolayer capacity), because the molecule has a large quadrupole moment. Buyanova et al. [104] therefore proposed the use of argon, the ω_m for this case being taken* as 15.4 Å^2. The measurement was accelerated by using an apparatus in which adsorption was performed on six specimens simultaneously, with independent preliminary evacuation of each specimen and with an integrator for more rapid and precise determination of the area under the desorption curve.

A correction must be made for the variation in flow rate with the temperature in the method of thermal desorption.

The adsorbent may be used as a finely divided powder to minimize the diffusion broadening of the peak [49, 98]. The curve may also be recorded as bulk speed (with a detector sensitive to the flow rate) against time, since the speed is affected by the adsorption and desorption of considerable volumes of gas. The concentration detector is replaced by a flow-rate detector, which need not be calibrated.

Kuge and Yoshikawa [32] proposed another gas-chromatographic method. We have seen that the shape of the peak is dependent on the sample size. Transition from monolayer adsorption to polymolecular layers causes very complicated variation in peak shape with sample size. An isotherm convex to the θ axis (e.g., a Langmuir isotherm) causes t_R to decrease as the sample size increases, whereas one convex to the gas-concentration axis causes t_R to increase with the size (Fig. 60). If the isotherm is a combination of these, the peak may take the forms shown in Fig. 64d and e. All of these peaks have [32] been observed for a column containing $Cu(Py)_2(NO_3)_2$ [108] (Fig. 81). A kink occurs at the top of the peak at the transition to polymolecular adsorption; this point corresponds to point b on the second type of isotherm in Brunauer's classification [60] (Fig. 65).

The area covered by a close-packed monolayer is

$$s = \frac{V_R}{V_M} \cdot N_A \cdot \omega_m, \qquad (IV, 35)$$

---

* The value $\omega_{m \text{ Ar}} = 13.6 \text{ Å}^2$ has been suggested [106].

in which V_R (cm^3) is the volume corresponding to that layer, N_A is Avogadro's number, ω_m is the area per molecule, and V_M is the mole volume (cm^3) of the gas at the temperature of the column. An approximate value for ω_m is [32]

$$\omega_m = \frac{6}{\sqrt{3}} \left(\frac{M}{4 \sqrt{2} N_A \delta} \right)^{2/3},$$

(IV, 36)

in which δ is the density of the liquid at the adsorption temperature and M is molecular weight. The method [32] has been used [108] to measure s for Cu(Py)$_2$(NO$_3$)$_2$ with benzene, CCl$_4$, cyclohexane, and n-hexane as test substances. The position of the kink of Fig. 81 was found to be dependent on the flow rate of the carrier gas, on T, and on the nature of the compound; it occurred for flow rates above 20 cm^3/min, and (presumably) diffusion broadening masks the kink at lower rates. No kink was found if T was above the boiling point of the compound.

APPLICATION OF GAS CHROMATOGRAPHY
TO ADSORPTION KINETICS

Gas chromatography allows one to examine not only equilibria but also rates of adsorption, which are of considerable importance in practical chromatography and also when adsorbents are used as catalysts or to remove materials.

The adsorbate—adsorbent interactions are reflected in t_R and peak shape for small sample sizes. The width is governed by the various diffusion processes occurring between the moment of injection and the moment of exit. The kinetic features may be examined if we can separate the broadening due to adsorption from that due to diffusion. A precise mathematical treatment of this problem is very complicated, so it is best to reduce the diffusion broadening until it can be neglected [33, 34] by using compounds that are very weakly adsorbed at room temperature, such as inert gases or other low-boiling nonspecifically adsorbed compounds. Lengthwise and eddy diffusion may occur as for strongly adsorbed substances; but adsorption and desorption are then largely ruled out. The importance of the diffusion processes may be estimated by comparing t_R and peak shape for an almost unadsorbed compound with those for a fairly strongly adsorbed compound. It is convenient to inject a mixture of these two compounds.

Eberly [33, 34] has shown that external diffusion is generally negligible for short columns, so the calculation is then simplified.

A point of particular interest is derivation of adsorption equilibria from nonequilibrium chromatograms with allowance for diffusion and kinetic broadening. Grubner (see [109, 110] for literature and some results) has given preliminary consideration to this.

LITERATURE CITED

1. A. Quantes and G. Rijnders, in collection: Proceedings of the Second International Symposium in Amsterdam and the Conference on the Analysis of Mixtures of Volatile Substances in New York [Russian translation], Moscow, IL, 1961, p. 120.
2. A. B. Littlewood, C. S. G. Phillips, and D. T. Price, J. Chem. Soc., 1955, p. 148.
3. D. White and C. T. Cowan, Trans. Faraday Soc., 54:557 (1958).
4. A. Gianetto and M. Panetti, Ann. Chim., 50:1713 (1960).
5. M. Panetti and G. Musso, Ann. Chim., 52:472 (1962).
6. H. Mackle, R. G. Mayrick, and J. J. Rooney, Trans. Faraday Soc., 56:115 (1960).
7. A. Liberti, L. Conti, and V. Crescenzi, Nature, 178:1067 (1956).
8. C. S. G. Phillips and P. L. Timms, J. Chromatog., 5:131 (1961).
9. I. A. Revel'skii, R. I. Borodulina, and G. M. Sovakova, Neftekhimiya, 4:804 (1964).
10. J. C. Giddings and S. L. Spenser, J. Chem. Phys., 33:1579 (1960).
11. E. N. Fuller and J. C. Giddings, J. Gas Chromatog., 7:222 (1965).
12. E. Wicke, Angew. Chem., B19:15 (1947).
13. E. Cremer and F. Prior, Z. Electrochem., 55:66 (1951).
14. E. Cremer and R. Müller, Z. Electrochem., 55:217 (1951).
15. S. J. Gregg and R. Stock, in collection: Gas Chromatography 1958, D. H. Desty (ed.), London, 1958, p. 90.
16. P. Fejes, L. Czaran, and G. Schay, Magy. Kem. Folyoirat., 68:11 (1962).
17. S. A. Greene and H. Pust, J. Phys. Chem., 62:55 (1958).
18. C. G. Scott, in collection: Gas Chromatography 1962, M. van Swaay (ed.), London, 1962, p. 7.
19. H. W. Habgood and J. F. Hanlan, Can. J. Chem., 87:843 (1959).
20. R. S. Petrova, E. V. Khrapova, and K. D. Shcherbakova, in collection: Gas Chromatography 1962, M. van Swaay (ed.), London, 1962, p. 18.
21. A. V. Kiselev and Ya. I. Yashin, Neftekhimiya, 4:634 (1964).
22. A. V. Kiselev, Yu. L. Chemen'kova, and Ya. I. Yashin, Neftekhimiya, 5:589 (1965).
23. K. Arita, Y. Kuge, and Y. Yoshikawa, Bull. Chem. Soc. Japan, 38:632 (1965).
24. A. V. Kiselev, in collection: Gas Chromatography, Transactions of the Third All-Union Conference on Gas Chromatography, Dzerzhinsk, Izd. Dzerzhinsk. Fil. OKBA, 1966, p. 15.

25. L. D. Belyakova, A. V Kiselev, and N. V. Kovaleva, Anal Chem., 36:1517(1964).

26. L. D. Belyakova, A. V. Kiselev, and N. V. Kovaleva, Dokl. Akad. Nauk SSSR, 157:646 (1964).

27. E. Cremer, Z. Anal. Chem., 170:219 (1959).

28. F. Wolf and H. Bayer, Chem. Techn. 11:142 (1959).

29. A. V. Kiselev, R. S. Petrova, and K. D. Shcherbakova, Kinetika i Kataliz, 5:526 (1964).

30. G. A. Gaziev, M. I. Yanovskii, and V. V. Brazhnikov, Kinetika i Kataliz, 1:548 (1960).

31. R. Stock, Anal. Chem., 33:966 (1961).

32. Y. Kuge and Y. Yoshikawa, Bull. Chem. Soc. Japan, 38:948 (1965).

33. P. E. Eberly, J. Appl. Chem., 14:330 (1964).

34. P. E. Eberly and L. H. Spencer, Trans. Faraday Soc., 57:289 (1961).

35. É. I. Semenenko, S. É. Roginskii, and M. I. Yanovskii, Kinetika i Kataliz, 5:490 (1964).

36. H. W. Kohlschutter and W. Höppe, Z. Anal. Chem., 197:33 (1963).

37. E. Gil-Av and M. Y. Herzberg, Proc. Chem. Soc., 1961, p. 316.

38. N. M. Turkel'taub, A. A. Zhukhovitskii, and N. V. Porshneva, Zh. Prikl. Khim., 34:1946 (1961).

39. F. Baumann, A. E. Straus, and J. F. Johnson, J. Chromatog., 20:1 (1965).

40. M. R. James, J. C. Giddings, and H. Eyring, J. Phys. Chem., 69:2351 (1965).

41. E. Cremer and H. Huber, Angew. Chem., 73:461 (1961).

42. D. R. Owens, A. C. Hamlin, and T. R. Phillips, Nature, 201:901 (1964).

43. J. N. Wilson, J. Am. Chem. Soc., 62:1583 (1940).

44. J. Weis, J. Chem. Soc., 1943, p. 297.

45. D. de Voult, J. Am. Chem. Soc., 65:532 (1943).

46. E. Gluckauf, Discussions Faraday Soc., 7:199 (1949).

47. D. H. James and C. S. G. Phillips, J. Chem. Soc., 1954, p. 1066.

48. S. J. Gregg, The Surface Chemistry of Solids, Second edition, London, Chapman and Hall, 1961, p. 329; S. J. Gregg and K. S. W. Sing, Adsorption, Surface Area and Porosity, London, Academic Press, 1967.

49. R. Vespalec and O. Grubner, in collection: Gas Chromatographie 1965, Berlin, Akademie Verlag, 1965, p. 517.

50. P. Fejes, E. Fromm-Czaran, and G. Schay, Acta Chim. Acad. Sci. Hung., 33:87 (1962).

51. G. Schay, Theoretical Principles of Gas Chromatography [Russian translation], Moscow, IL, 1963, p. 136.

52. F. Fejes, E. Fromm-Czaran, and G. Schay, in collection: Gas Chromatographie 1961, H.-P. Angele and H. G. Struppe(eds.), Berlin, Akademie Verlag, 1961, p. 42.

53. S. Z. Roginskii, M. I. Yanovskii, Lu Pei-chang, G. A. Gaziev, G. M. Zhabrova, B. M. Kodenatsi, and V. V. Brazhnikov, Dokl. Akad. Nauk SSSR, 133:878(1960).

54. N. A. Shilov, L. K. Lepin', and S. A. Voznesenskii, Zh. Russk. Fiz.-Khim. Obshch., 51:1107 (1927).

55. A. A. Zhukhovitskii, Ya. B. Zabezhinskii, and D. S. Sominskii, Zh. Fiz. Khim., 13:303 (1939).

56. A. A. Zhukhovitskii and N. M. Turkel'taub, in collection: Gas Chromatography, Moscow, Gostoptekhizdat, 1962.

57. H. F. Huber and A. J. M. Keulemans, in collection: Gas Chromatography, M. van Swaay (ed.), London, 1962, p. 26.

58. Ya. I. Gerasimov, V. P. Dreving, E. N. Eremin, A. V. Kiselev, V. P. Lebedev, G. M. Panchenkov, and A. I. Shlygin, in: Textbook of Physical Chemistry Vol. 1, Ya. I. Gerasimova (ed.), Moscow, Goskhimizdat, 1963, p. 543.

59. L. D. Belyakova, A. V. Kiselev, and N. V. Kovaleva, Bull. Soc. Chim. France, No. 1:285 (1967).

60. S. Brunauer, Adsorption of Gases and Vapors, Princeton University Press, 1945.

61. S. Brunauer, P. H. Emmett, and E. Teller, J. Am. Chem. Soc., 60:309 (1938).

62. A. V. Kiselev, Discussion Faraday Soc., 40:205 (1965).

63. E. Cremer and H. F. Huber, in collection: Gas Chromatography, Instr. Soc. Am. Symposium, N. Brenner (ed.), New York, Academic Press, 1962, p. 169.

64. P. J. Kipping and D. G. Winter, Nature, 205:1002 (1965).

65. R. A. Beebe, P. L. Evans, T. C. W. Kleinsteuber, and L. W. Richards, J. Phys. Chem., 70:1009 (1966).

66. A. Littlewood, in: Gas Chromatography, Principles, Techniques, and Applications, New York, Academic Press, 1962.

67. R. M. Barrer and L. V. C. Rees, Trans. Faraday Soc., 57:999 (1961).

68. B. G. Aristov, V. L. Keibal, A. V. Kiselev, and K. D. Shcherbakova, in collection: Gas Chromatography, Moscow, Izd. NIITEKhIM, Issue 6, 1967, p. 61.

69. D. H. Everett, Trans. Faraday Soc., 46:957 (1950)-

70. L. D. Belyakova, A. V. Kiselev, and N. V. Kovaleva, Zh. Fiz. Khim., 40:1494 (1966).

71. A. V. Kiselev, I. A. Migunova, and Ya. I. Yashin, Zh. Fiz. Khim., 42:1235 (1968).

72. J. A. Barker and D. H. Everett, Trans. Faraday Soc., 58:1608 (1962).

73. A. V. Kiselev and D. P. Poshkus, Zh. Fiz. Khim., 41:2647 (1967).

74. A. B. Littlewood, Anal. Chem., 36:1441 (1964).

75. V. L. Keibal, A. V. Kiselev, I. M. Savinov, V. L. Khudyakov, K. D. Shcherbakova, and Ya. I. Yashin, Zh. Fiz. Khim., 41:2234 (1967).

76. J. J. Carberry, Nature, 189:391 (1961).

77. R. L. Gale and R. A. Beebe, J. Phys. Chem., 68:555 (1964).

78. A. A. Isirikyan and A. V. Kiselev, Zh. Fiz. Khim., 36:1164 (1962).

79. A. V. Kiselev, Yu. S. Nikitin, R. S. Petrova, K. D. Shcherbakov, and Ya. I. Yashin, Anal. Chem., 36:1526 (1964).

80. A. V. Kiselev, Yu. S. Nikitin, R. S. Petrova, and Fan Ngok Than, Kolloidn. Zh., 27:368 (1965).

81. I. V. Kolosnitsyna and R. S. Petrova, Kolloidn. Zh., Vol. 29:815 (1967).

82. A. V. Kiselev and Ya. I. Yashin, in collection: Gas Chromatography, Transactions of the Third All-Union Conference on Gas Chromatography, Dzerzhinsk, Izd. Dzerzhinsk. Fil. OKBA, 1966, p. 181.

83. G. I. Berezin, A. V. Kiselev, and V. A. Sinitsyn, Zh. Fiz. Khim., 41:926 (1967).

84. G. I. Berezin, A. V. Kiselev, and A. A. Kozlov, Zh. Fiz. Khim., 41:1426
     (1967).
85. R. Aubeau, J. Leroy, and L. Champeix, J. Chromatog., 19:249 (1965).
86. H. W. Habgood, Can. J. Chem., 42:2340 (1964).
87. A. A. Kubasov, I. V. Smirnova, and K. V. Topchieva, Kinetika i Kataliz,
     5:520 (1964).
88. R. D. Oldenkomp and G. Houghton, J. Phys. Chem., 67:303 (1963).
89. S. Skornik, M. Steinberg, and P. S. Stone, Israel J. Chem., 1:320 (1963).
90. S. Ross, J. K. Saelens, and J. P. Olivier, J. Phys. Chem., 66:696 (1962).
91. R. C. Chirnside and C. G. Pope, J. Phys. Chem., 68:2377 (1964).
92. A. V. Kiselev, G. M. Petov, and K. D. Shcherbakova, Zh. Fiz. Khim., 41:1418
     (1967); G. M. Petov and K. D. Shcherbakova, in collection: Gas Chromato-
     graphy 1966, A. B. Littlewood (ed.), London, Institute of Petroleum, 1967,
     p. 50.
93. H. Knozinger and H. Spannheimer, J. Chromatog., 16:1 (1964).
94. N. N. Avgul', G. I. Berezin, A. V. Kiselev, and I. A. Lygina, Izv. Akad. Nauk
     SSSR, Otd. Khim. Nauk, 1961, p. 205.
95. N. N. Avgul', A. V. Kiselev, and I. A. Lygina, Kolloidn. Zh., 23:369 (1961).
96. N. N. Avgul', A. V. Kiselev, and I. A. Lygina, Kolloidn. Zh., 23:513 (1961).
97. F. M. Nelson and F. T. Eggertson, Anal. Chem., 30:1387 (1958).
98. O. G. Grubner, Z. Phys. Chem., 216:287 (1961); E. Smolkova, V. Kristofikova,
     L. Feltl, and O. Grubner, Collection Czech. Chem. Commun., 31:450 (1966).
99. K. V. Wise and E. H. Lee, Anal. Chem., 34:301 (1962).
100. H. W. Daeschner and F. H. Stross, Anal. Chem., 34:1150 (1962).
101. L. S. Ettre, N. Brenner, and E. W. Cieplinski, Z. Phys. Chem., 219:17 (1962).
102. A. J. Holly, J. Appl. Chem., 13:392 (1963).
103. H. Atkins, Anal. Chem., 36:579 (1964).
104. N. E. Buyanova, G. B. Gudkova, and A. P. Karnaukhov, Kinetika i Kataliz,
     6:1085 (1965).
105. T. B. Gavrilova and A. V. Kiselev, Zh. Fiz. Khim., 39:2582 (1965).
106. B. G. Aristov and A. V. Kiselev, Zh. Fiz. Khim., 37:2502 (1963); 38:18
     (1964).
107. G. Schay, in collection: Gas Chromatographie 1965, H. G. Struppe (ed.),
     Berlin, Akademie Verlag, 1965.
108. A. G. Altenau and L. B. Rogers, Anal. Chem., 37:1432 (1965).
109. O. Grubner and E. Kučera, in collection: Gas Chromatographie 1965, H. G.
     Struppe (ed.), Berlin, Akademie Verlag, 1965, p. 157.
110. E. Kučera, M. Ralek, and A. Zikánová, Collection Czech. Chem. Commun.,
     31:852 (1966).
111. A. V. Kiselev, Paper presented at ACS meeting, San Francisco, April 1968.
112. S. P. Dzhavadov, A. V. Kiselev, and Yu. S. Nikitin, Zh. Fiz. Khim., 41:1131
     (1967).
113. A. V. Kiselev, in collection: Advances in Chromatography, Vol. 4, J. C.
     Giddings and R. A. Keller (eds.), New York, Dekker, 1967, p. 113.
114. V. L. Keibal, A. V. Kiselev, I. M. Savinov, V. L. Khudyakov, K. D. Shcher-
     bakova, and Ya. I. Yashin, Zh. Fiz. Khim., 41:2234 (1967).

115. E. Eberly, J. Phys. Chem., 65:68 (1961).
116. R. M. Barrer and R. M. Gibbons, Trans. Faraday Soc., 61:948 (1965).
117. R. M. Barrer and B. Coughlan, in collection: Molecular Sieves, Soc. Chem. Ind.,
      London, pp. 233, 241.
118. S. S. Khvoshchev, S. P. Zhdanov, G. M. Belotsevkovskii, and V. I. Redin,
      Zh. Fiz. Khim., 42:171 (1968).
119. L. Bertsh and H. W. Habgood, J. Phys. Chem., 67:1621 (1963).
120. J. W. Ward and H. W. Habgood, J. Phys. Chem., 70:1178 (1966).
121. N. K. Bebris, A. V. Kiselev, and Yu. S. Nikitin, Neftekhimiya, No. 9 (1969).
122. I. Yu. Babkin and A. V. Kiselev, Dokl. Akad. Nauk SSSR, 129:357 (1959); Zh.
      Fiz. Khim., 36:2448 (1962); I. Ya. Babkin, A. V. Kiselev, and A. Ya. Kovolev,
      Dokl. Akad. Nauk SSSR, 136:373 (1964).
123. A. V. Kiselev, Discussions Faraday Soc., 40:205 (1965).
124. C. Vidal-Madjar and G. Guiochon, Separation Sci., 2:155 (1967).
125. T. N. Gvozdovich, A. V. Kiselev, and Ya. I. Yashin, Chromatographia, 6:234
      (1969).
126. A. A. Isirikjan and A. V. Kiselev, J. Phys. Chem., 65:601 (1961).
127. G. I. Berezin, A. V. Kiselev, and Yu. S. Nikitin, Kinetika i Kataliz, 8:238
      (1967).
128. S. P. Dzhavadov, A. V. Kiselev, and Yu. S. Nikitin, Ibid., 8:238 (1967).
129. V. I. Kalmanovskii, A. V. Kiselev, G. G. Sheshenina, and Ya. I. Yashin, Gas
      Chromatography, Moscow, Izd. NIITEKhIM, Issue 6, 1967, p. 45.
130. L. D. Belyakova, A. V. Kiselev, and K. D. Shcherbakova, Kolloidn. Zh.,
      31:18 (1969).

Chapter V

Analytical Use of Gas-Adsorption Chromatography

ANALYSIS OF HYDROGEN
ISOTOPES AND ISOMERS

Gas adsorption has attracted much attention for these operations, which are of importance in biochemistry [1] and geophysics [2]. The problem is complicated by two equilibria:

$$\text{p-}H_2 \rightleftarrows o\text{-}H_2 \quad \text{and} \quad H_2 + D_2 \rightleftarrows HD.$$

Ordinary hydrogen consists of three parts o-H_2 and one part p-H_2, the rate of o-p conversion being very small, though it becomes appreciable on iron oxide. The second reaction is catalyzed by heat and even by the column packing. Palladium black shifts the equilibrium to the left.

Martin and Synge [3] in 1941 proposed chromatography for isotope separation. Glueckauf and Kitt [4] were the first to make pure D_2 by frontal chromatography on palladium black mounted on asbestos. Thomas and Smith [5] used this on quartz to separate H_2 and D_2 by development, but obtained only partial separation.

The simplest method of analysis for deuterium in the presence of hydrogen is to use ordinary hydrogen as the carrier gas, with measurement of the heights of the partially resolved HD and D_2 peaks [6-9]; activated charcoal or CaA zeolite may be used at room temperature.

In any case, HD may be transformed to H_2 and D_2 by adding palladium black to the column. The thermal conductivity of D_2 dif-

146

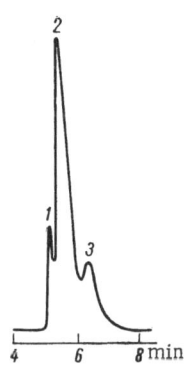

Fig. 82. Separation of hydrogen isotopes and isomers at −196°C on a 360 × 0.6 cm column of active alumina (He 100 ml/min, katharometer): 1) p-H_2; 2) o-H_2 + HD; 3) D_2.

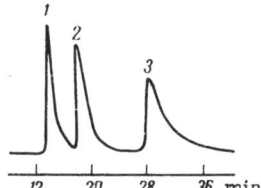

Fig. 83. Separation of hydrogen isotopes on a 180 × 0.4 cm alumina column −196°C (He 100 ml/min): 1) H_2; 2) HD; 3) D_2.

fers from that of H_2 sufficiently for a katharometer to detect 2% deuterium in hydrogen.

The isomers o-H_2 and p-H_2 are well separated [9] on an active-alumina column at 77°K with He as carrier gas (Fig. 82), with incomplete separation of o-D_2 and p-D_2. Quantitative analysis of H_2, HD, and D_2 is often performed with active alumina containing a little iron oxide [10, 11] or chromium oxide [12, 13] at 77°K with He or Ne as carrier (Fig. 83). The iron (chromium) oxide prevents the separate elution of ortho and para forms of H_2 and D_2, as the mutual conversion of these is much more rapid than the passage of the gas along the column, so t_R is the mean for the two forms. Accuracy 0.2%. The limits of detection for HD and D_2 are less than 0.01%, so the natural abundance of deuterium can be detected. The symmetry of the peaks is very dependent on the degree of activation; strong activation causes the D_2 peak to become unsymmetrical. Pronounced deactivation with water vapor gives symmetrical peaks, but these partly overlap. Partial deactivation gives complete separation of H_2, HD, and D_2.

Only two columns used together provide analysis of all mixtures of p-H_2, o-H_2, HD, and D_2 in one operation [14, 15]. The first column contains alumina or a molecular sieve (NaX zeolite), while the second contains alumina with a little iron oxide. The isomers are separated on the first column, while the second maintains o-p equilibrium but separates the isotopes (Fig. 84) [16-23]. The Q are [24] (in kcal/mole) 1.4 for p-H_2, 1.55 for o-H_2, 1.51 for HD, 1.58 for o-D_2, and 1.65 for p-D_2. Even in spite of these small differences, the separation is quite feasible.

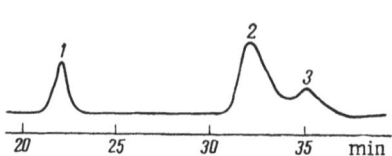

Fig. 84. Separation of hydrogen isotopes on a 200 × 0.4 cm NaX column at −195°C: 1) HD; 2) o-D_2; 3) p-D_2.

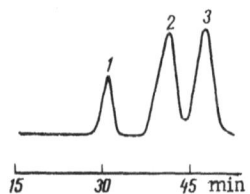

Fig. 85. Separation of hydrogen isotopes on a 2.4 m alumina column at 77°K (carrier neon): 1) HT; 2) DT; 3) T_2.

Fig. 86. Separation of hydrogen isotopes and isomers at −125°C on a glass capillary column 80 m × 0.27 mm with a porous surface (carrier neon): 1) He; 2) p-H_2; 3) o-H_2; 4) HD; 5) o-D_2; 6) p-D_2.

Smith and Carter [25] were the first to report separation of H_2, HT, and T_2 in a mixture containing 1% tritium on a zeolite column at −160°C with helium as carrier gas. This was then extended [26, 27] to traces (to 10^{-4} M) of tritium at 77°K, and finally to trace amounts in a complete mixture (with a somewhat modified technique) [27] (Fig. 85). The effects of carrier gas, adsorbent, and activation conditions on the separation were also examined. The detector was a flow ionization chamber joined in series with a thermistor katharometer [27].

Excellent results have been reported by Mohnke and Saffert [28] for a glass capillary column working by gass adsorption.

Detailed studies of these separations have been made by White and Lassettre [29] and Evett [30]. Phillips et al. [31] showed that low-temperature adsorption on CaA (5A) molecular sieve, followed by desorption, can provide a preliminary separation of deuterium from hydrogen. Recent detailed reviews of the subject are by Sako-dynskii [32], Akhtar and Smith [33], and Cercy and Botter [34].

ANALYSIS OF INORGANIC GASES AND METHANE

The isotherms for microporous adsorbents are virtually linear at the concentrations used in gas chromatography, so adsorption methods are widely used in gas analysis, the commonest adsorbents being charcoal, molecular sieves (CaA and NaX zeolites), and microporous silica and alumina gels [35-37]; microporous glasses have also been used [38, 39].

Gases are separated because their Henry constants differ. Most gases are nonpolar and so are adsorbed mainly via nonspecific dispersion forces (except for compounds such as CO or N_2, which have high dipole or quadrupole moments), so adequate differences in uptake require the use of adsorbents with very small pores, since the separation of nonpolar gases is controlled more by the specific surface s than by the chemical nature of the surface. This means that separations are usually obtainable with nonspecific adsorbents (activated charcoal) and with specific hydroxylated ones (silica and alumina gels), or with cationized ones (zeolites) provided that s is adequate. Other things being equal, the separation is the better the narrower the size range of the pores.

Formerly, charcoal was used to separate the inert gases, H_2, O_2, N_2, CO, and CH_4 [40]; but O_2 and N_2 are separated only on very long columns, so artificial zeolites have now become widely used for this purpose [41], the separation on these being good even at 100°C [42]. CaA and NaX give good separation of mixtures of H_2, O_2, N_2, CH_4, and CO, but CO_2 is desorbed only above 150°C [44].

Polar compounds (H_2S, NH_3, CS_2, NO_2, etc.) are adsorbed very strongly by zeolites [45], which are therefore unsuitable for separating these gases, though they are widely used to remove them from gas mixtures.

O_2-N_2 mixtures have [46] been separated on manganese oxide at high T.

Greene and Pust [47] used a charcoal column to separated a mixture containing H_2, O_2 N_2, CO, CH_4, and CO_2 by the use of programmed temperature control from room temperature up to 170°C.

Argon and oxygen are similar in polarizability, so these are difficult to separate on all adsorbents. They have been analyzed [48] only from the difference in two detectors, with argon or oxy-

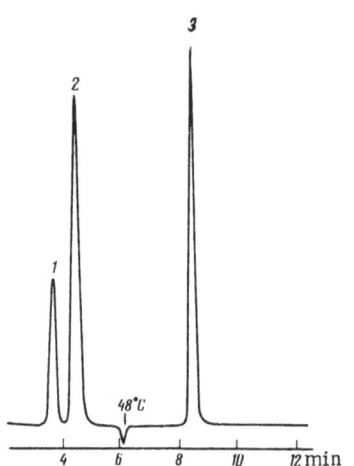

Fig. 87. Separation of: 1) Ar; 2) O_2; 3) N_2 on CaA, with separation of Ar and O_2 at −40°C, the column then being heated to +48°C.

gen used in turn as carrier. A 10 m column of CaA separates Ar and O_2 at room temperature [49]. Ar and O_2 may be separated by the use of cooled columns or by catalytic conversion of oxygen to H_2O, with hydrogen as carrier gas; if N_2 is present, the analysis is performed in two stages. For instance, the O_2 and Ar are separated at −72°C on CaA [43], and then N_2 is separated from the Ar−O_2 mixture at 25°C. Heylmun [50] has used CaA as −72°C to determine traces of O_2 in pure Ar. Hydrogen has been used as carrier [51] with catalytic conversion of O_2 to H_2O for room-temperature determination of Ar. Abel [52] and Swinnerton et al.[53] converted this water to acetylene by reaction with CaC_2 and separated this from He and Ar on silica gel. Ar−O_2−N_2 mixtures from air have been separated in a single run on CaA [54]; the column was cooled to −40° to separate Ar and O_2, and then the N_2 was eluted by raising the temperature to 48° (Fig. 87). Ar and oxygen may be separated completely on CaA zeolite dried at 400°C [252, 253].

Detailed studies have been made [55, 56] of the separation of He, O_2, N_2, CH_4, and CO by zeolites of various types (Fig. 88).

CO is well separated from O_2, N_2, and CH_4 by zeolites and follows methane; whereas it emerges before methane on alumina gel, silica gel, and charcoal [57]. CO_2 is adsorbed strongly by zeolites and is almost immobile at room temperature, though it is eluted fairly readily from silica gel. Program temperature control with CaA gives (Fig. 89) rapid separation of low-boiling gases and also CO_2 [58].

He, O_2, N_2, CO, and CH_4 are well separated by porous glass (Fig. 90) [59]; as for silica gel, N_2 and O_2 are not separated, while CO runs ahead of CH_4.

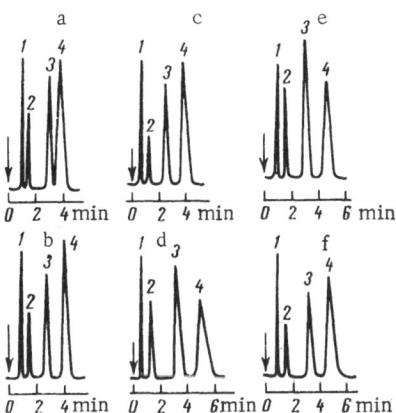

Fig. 88. Separation of: 1) He; 2) O_2; 3) N_2; 4) CH_4 at 20°C on 100 × 0.45 cm columns of various zeolites (hydrogen 50 ml/min; 2 ml sample): a) VNIINP CaX; b) VNIINP NaX; c) Linde 13X; d) VNIINP CaA; e) as previous, spherical form; f) Linde 5A.

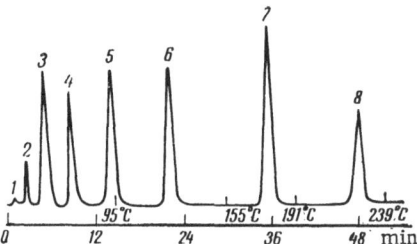

Fig. 89. Separation of low-boiling gases on a 122 × 0.63 cm CaA column with temperature rising at 4 deg/min (argon 60 ml/min): 1) H_2; 2) O_2; 3) N_2; 4) CH_4; 5) CO; 6) C_2H_6; 7) CO_2; 8) C_2H_4.

The principal components of air (oxygen, nitrogen, CO_2) may be separated simultaneously on a column of Microtek alumina gel at 0°C after preliminary treatment with CO_2 [254].

H_2, N_2, CO, and CO_2 have been separated on silica gel at 35°C [60].

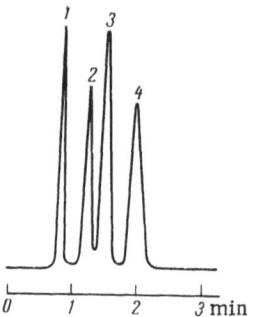

Fig. 90. Separation on 100 × 0.4 cm porous-glass column (pore diameter about 10 Å) at 20°C (hydrogen 30 ml/min, grain size of glass 0.05-0.2 mm): 1) He; 2) $O_2 + N_2$; 3) CO; 4) CH_4.

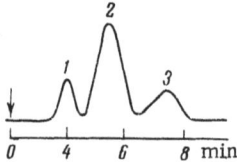

Fig. 91. Separation of: 1) N_2; 2) Kr; 3) CH_4 at 20°C on a column with activated charcoal and CaA (ratio of lengths of layers 1 : 10, grain size 0.2-0.4 mm, hydrogen 2.8 ml/sec).

Janak [61, 62] has discussed the simultaneous separation of all low-boiling gases; he showed that it possible to separate mixtures of the inert gases with H_2, CO, and CH_4 [62], and he determined V_R for He, Ne, Ar, Kr, CH_4, and Xe on charcoal at 20° [62]. Here He and Ne are not separated, while Xe emerges after a long time and O_2 and N_2 are not separated from Ar. Greene [63] also made a detailed study of the separation of mixtures of inert gases with O_2, N_2, and CH_4 on charcoal at $-196°C$, on silica gel at 23°C, and with molecular sieves at 23°C. He and Ne are not separated on charcoal at room temperature, but are at $-196°C$. Ar, Kr, and Xe are readily separated at room temperature on silica gel, but CH_4 (if present) coincides with Kr; separation of Ar, Kr, CH_4, and Xe requires the use of molecular sieves. Complete elution of Xe requires the column to be heated to 100°C.

N_2, Kr, and CH_4 may be separated at room temperature on a column containing 1 part charcoal and 10 parts CaA [64] (Fig. 91). H_2 is poorly separated from He and Ne on charcoal at room temperature [61] but is separated well on CaA at $-78°C$. Janak separated H_2 and He on Celite coated with palladium [65].

He and Ne in natural gas have been determined by gas chromatography [66, 67]. The amounts of N_2, Ar, and He in natural gas provide important geochemical evidence, especially on the geological age of natural gas.

Complete analyses for He, Ne, and H_2 in air have been performed with SKT charcoal at 15°C with Ar as carrier [68, 69] (Fig.

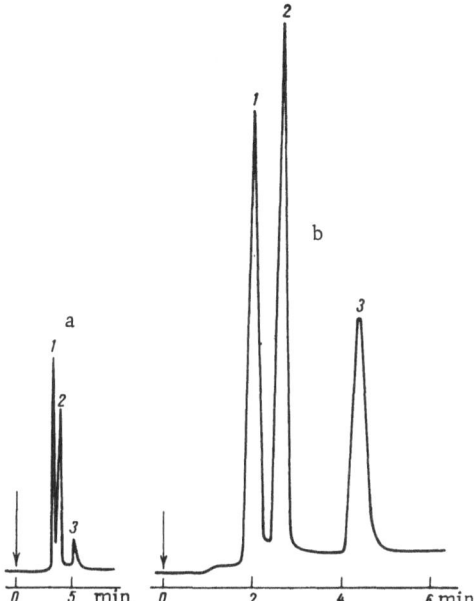

Fig. 92. Separation of: 1) He; 2) Ne; 3) H_2 at 25°C
on activated charcoal: a) 530 × 0.45 cm column of
SKT charcoal, grain size 0.25-0.5 mm, argon 26
ml/min; b) 400 × 0.4 cm column of Saran charcoal, grain size 0.25-0.5 mm.

92a) and Saran charcoal (Fig. 92b), the air being pretreated by
frontal adsorption to concentrate these components. The results
were 0.00054% He, 0.002% Ne, and 0.00028% H_2 in street air
(Fig. 92).

Complete analysis of the atmosphere is possible; He, H_2, Ne,
Ar, Kr, Xe, O_2, N_2, CO, and CO_2 in air have been determined [70]
with activated charcoal (H_2, Ne, He), silica gel (Ar, Kr, Xe, and
CO_2), and CaA (O_2, N_2, CO). It was found that a 5 ml sample contained 0.028% CO_2, 0.2-1% Ar, 20.2% O_2, and 78.7% N_2.

Yashin [71] has examined various Russian charcoals (SKT,
AG-3, AR-3, BAU) for separation as regards light gases; SKT was
the best (Fig. 93).

Glueckauf and Kitt [72] have assayed atmospheric Kr and Xe by
gas chromatography. Gas-chromatographic analysis of inert gases

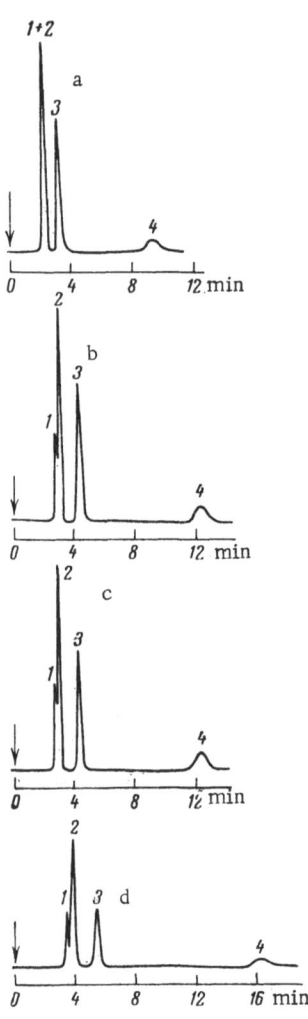

Fig. 93. Chromatograms of: 1) O_2; 2) N_2; 3) CO; 4) CH_4 at 30°C on 200 × 0.5 cm columns with different activated charcoals (0.5 ml sample, He 60 ml/min, grain size 0.25-0.5 mm): a) BAU; b) AR-3; c) AG-3; 4) SKT.

is widely used in air-liquefying plants, especially for determining Kr and Xe in the primary krypton component [73] and in technical Kr [74], as well as for Kr and Xe in the first oxygen concentrate [75].

Dynamic-temperature and chromothermographic techniques can be used to advantage with inert gases [76, 77].

Inorganic gases containing S and F may be separated [18, 78, 79]; silica gel has been used [78] to separate CO_2, COS, H_2S, CS_2, and SO_2 (Fig. 94). A composite column with silica gel and Teflon bearing a liquid phase will [79] separate O_2, CF_2O, and CF_4. A mixture of CO_2 and CF_2O is well separated by a composite column consisting of 60 cm of Porapak T and 120 cm of Porapak N [255].

Analytical applications exist in various branches of science and technology.

A common problem is determination of traces of the permanent gases in pure gases, as in the determination of inorganic gases in electrolytic chlorine [80], nitrogen in the Ar used as an inert medium in semiconductor manufacture [81], or O_2, N_2, CO, and CH_4 in pure ethylene [82], inorganic gases in CO_2 used as a coolant in nuclear reactors [83], and contaminants in pure He [84]. Other examples are the assay of CH_4 in air in pits [85] and of hydrogen in mine gases [86].

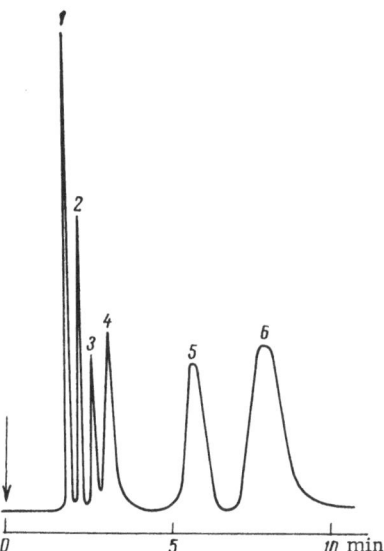

Fig. 94. Chromatogram given by a 30 cm
silica gel column at 100°C (He 40 ml/min):
1) $O_2 + N_2$; 2) CO_2; 3) COS; 4) H_2S; 5) CS_2;
6) SO_2.

Quantitative assay of components is often needed; e.g., hydrogen in water indicates corrosion in high-pressure boilers [87],
the composition of combustion products indicates the completeness
of combustion in boilers [88], and analysis of effluent gases serves
to monitor various processes [89]. Assay of Ar and CH_4 in the cir-
culating $N_2 - H_2$ mixture is used in the production of synthetic ammonia [90] and so on.

Gas chromatography is also being used in metallurgy to determine gases dissolved in metals, especially inorganic gases ex-
tracted by vacuum melting. Zeolites are often used as the adsorbents. H_2 and CO have been assayed in alloys of iron, uranium, and
zinc [91], in steel [92], and in copper and aluminum alloys, while
nitrogen and hydrogen in aluminum have been measured [93]. Methods have been described [94] for analysis of gases in rocks and
minerals.

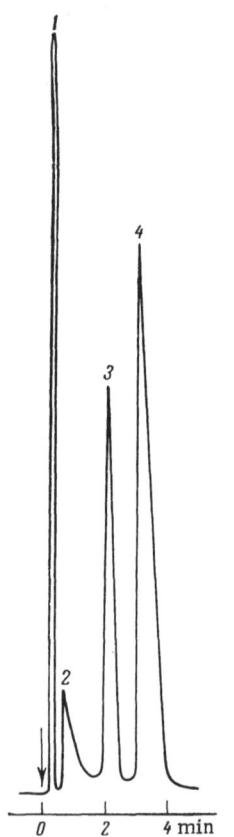

Fig. 95. Chromatogram from
200 × 0.3 cm column of SKT
charcoal at 70°C (He 60 ml/min):
1) $O_2 + N_2$; 2) NO; 3) CO_2; 4) N_2O.

Dissolved oxygen has been assayed for lubricating oils, petroleum, and various fractions of crude petroleum [95, 96]; dissolved gases have been measured in water [97], dissolved CO_2 in wine [98], and dissolved hydrogen in beer [99].

Gas-adsorption chromatography has become widely used in biochemistry; analysis of air has been used [100] in research on respiration, photosynthesis, and nitrogen fixation. Ar in air (up to 1%) may also be determined, although this is inert and so is less important to biochemistry. The method has been applied to blood gases, in particular CO [101], gases in biological fluids [102], gases in soils and fertilizers, and decomposition products from organic compounds [103]. In the last case the products include N_2O, NO, NO_2, NH_3, and H_2S, whose separation and analysis by gas chromatography represents a difficult problem. CO_2 and N_2O are not separated on silica gel, but they can be separated on charcoal [104].

N_2O and CO_2 have [105] been separated on composite columns. The first column (length 20 cm, 180°C) contained NaX zeolite, while the second contained silica gel. CO_2 mixed with CO may be analyzed on a molecular sieve bearing water by quantitative conversion of the CO_2 to NO and analysis for NO, which is retained 1.5 times more strongly than N_2 by the zeolite [106].

N_2O, NO, N_2, CO, CO_2, and H_2 are produced in the combustion of explosives; these have been separated below 0°C on specially treated silica gel [107]. The oxides of nitrogen may be separated

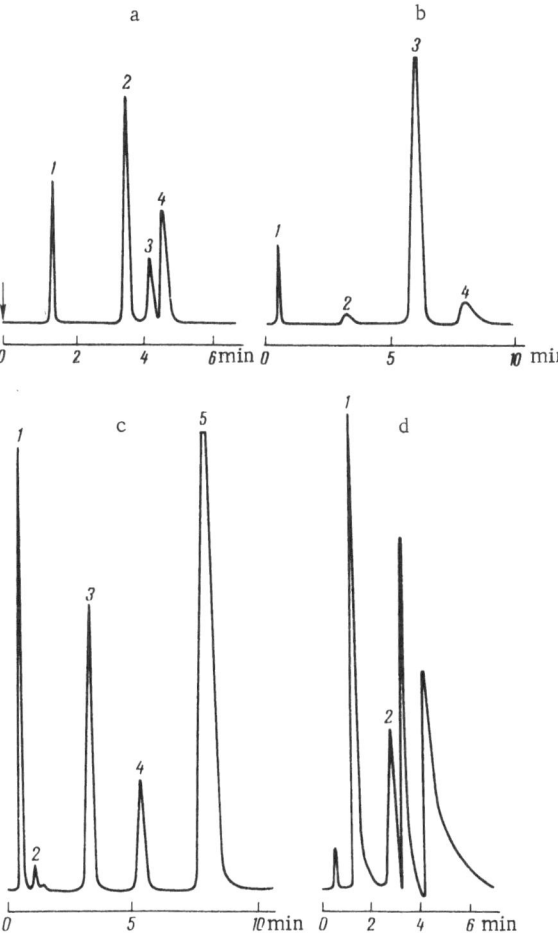

Fig. 96. Chromatograms given by 180 × 0.4 cm column of porous polystyrene (s = 600 m^2/g, H$_2$ 90 ml/min): a): 1) He; 2) N$_2$; 3) O$_2$; 4) Ar at −78°C; b): 1) air; 2) H$_2$O; 3) HCN; 4) SO$_2$ at 70°C; c): 1) O$_2$+N$_2$; 2) CO$_2$; 3) H$_2$S; 4) COS; 5) SO$_2$ at 68°C; d): 1) NH$_3$; 2) H$_2$O at 130°C (sensitivity altered in recording water peak).

on SKT charcoal [108] (Fig. 95); these components emerge in order of increasing boiling point from a porous polymer [109]. This adsorvent also provides good separation of N$_2$, O$_2$, and Ar (Fig. 96a), air, water, HCN, and SO$_2$ (Fig. 96b), air, CO$_2$, H$_2$S, COS, and SO$_2$ (Fig. 96c), and NH$_3$ and H$_2$O (Fig. 96d).

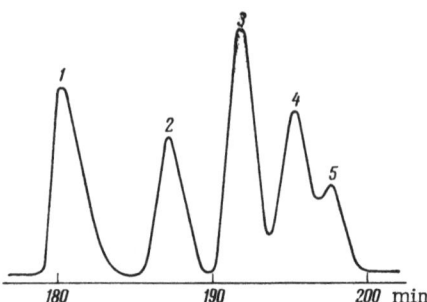

Fig. 97. Separation of deuteromethanes at
−188°C on a porous glass capillary column
35 × 0.3 mm (55,000 theoretical plates):
1) CH_4; 2) CH_3D; 3) CH_2D_2; 4) CHD_3; 5) CD_4.

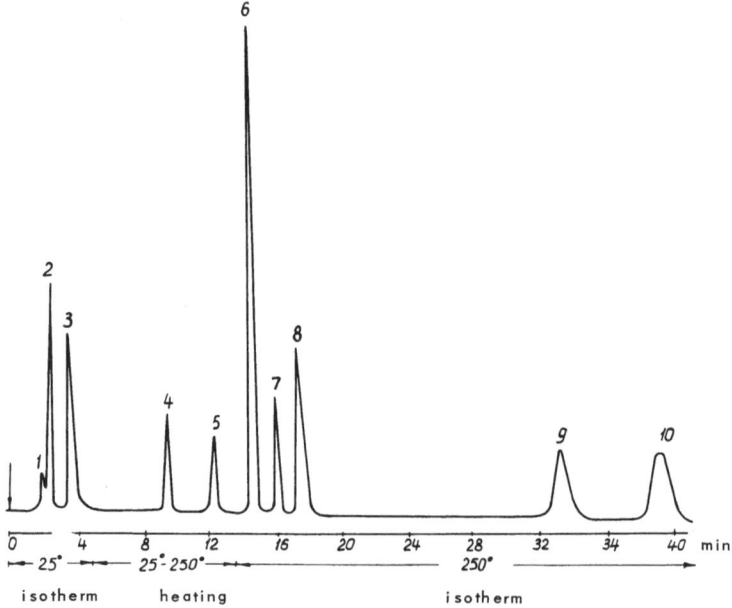

Fig. 98. Chromatogram on a 300 × 0.3 cm Saran charcoal column (He 30
ml/min, katharometer) for: 1) oxygen; 2) nitrogen; 3) CO; 4) methane;
5) actylene; 6) CO_2; 7) ethylene; 8) ethane; 9) propene; 10) propane.

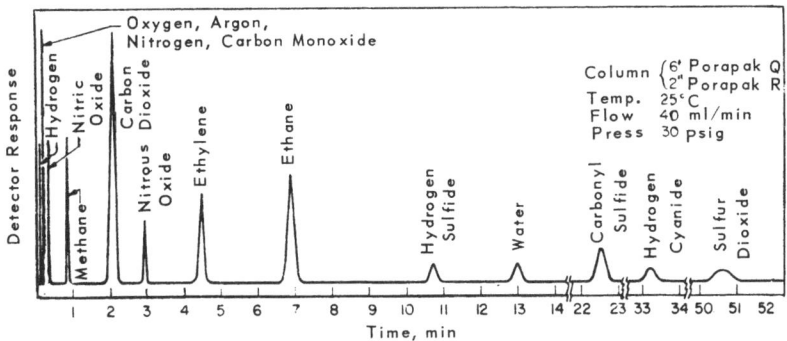

Fig. 99. Chromatogram of a mixture of gases on porous polymers, conductors given in the figure.

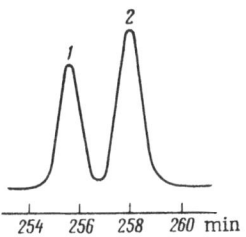

Fig. 100. Separation of oxygen isotopes O_2^{16} (1) and O_2^{18} (2) at 77°K on a glass adsorption column 175 cm × 0.3 mm (350,000 theoretical plates).

A mixture of low-boiling inorganic gases and light hydrocarbons may be separated on a single column of Saran charcoal (Fig. 98) [256]. A mixture of low-boiling inorganic gases, nitrogen, oxides, sulfur compounds, and light hydrocarbons may be separated on a porous-polymer column with programmed temperature control (Fig. 99) [257].

The deuteromethanes (CH_4, CH_3D, CH_2D_2, CHD_3, CD_4) have [110] been completely separated on a glass capillary column 35 m × 0.3 mm (55,000 theoretical plates) (Fig. 97), as have the isotopes of oxygen (Fig. 100).

NO_2 at the ppm level in N_2 and O_2 has been determined with an electron-capture detector [111], and traces of Ar, H_2, CO_2, O_2, N_2, and Ne have been determined with a radioactive ionization detector [112].

ANALYSIS OF HYDROCARBON GASES

Light hydrocarbons are usually analyzed by gas adsorption, because there are very few liquid phases that provide good separation, especially for ethane and ethylene [113], so very long columns are used, these being operated below 40°C, on account of the

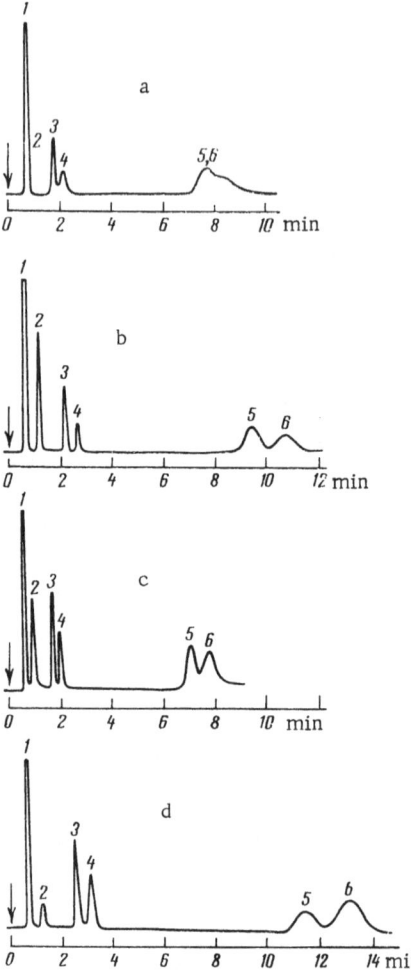

Fig. 101. Mixture of: 1) $O_2 + N_2$; 2) CH_4; 3) C_2H_4; 4) C_2H_6; 5) C_3H_6; and 6) C_3H_8 at 228°C on a 200×0.5 cm column (He 60 ml/min) with activated charcoals: a) BAU; b) AR-3; c) AG-3; d) SKT.

high volatility of the liquid phases [114]. Gas–adsorption analysis is done mainly with silica and alumina gels, which separate C_1–C_3 in comparatively short times at moderate temperatures, whereas separation of C_4 and C_5 requires the column to be heated to 100° [115] or above [116]. The unsaturated hydrocarbons are usually

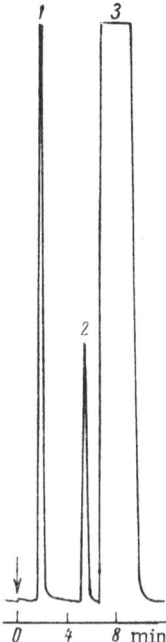

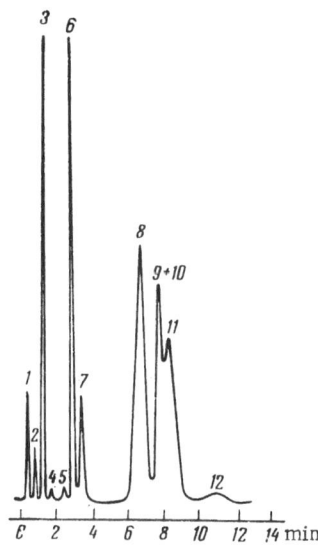

Fig. 102. Mixture of hydrocarbons at 130°C on a 100 ×0.4 cm column of SKT activated charcoal (flame-ionization detector, N_2 40 ml/min): 1) methane; 2) acetylene; 3) ethylene.

Fig. 103. Hydrocarbons at 60°C on a 130 × 0.4 cm column of macroporous silica gel modified with 10% KOH (N_2 38 ml/min, flame-ionization detector): 1) methane; 2) ethane; 3) ethylene; 4) propane; 5) propylene; 6) isobutane; 7) n-butane; 8) α-butylene; 9) trans-butylene; 10) cis-butylene; 11) isobutylene; 12) divinyl.

desorbed later than the saturated ones, on account of the additional specific interaction of the π-bonds with the surface OH groups. The uptake and selectivity are much reduced if these adsorbents are loaded with water; the effects for silica gels have been considered in several papers [117].

Activated charcoals are also used to separate light hydrocarbons; the nonspecificity causes the hydrocarbons to emerge in order of boiling point (CH_4, C_2H_2, C_2H_4, C_2H_6, C_3H_6, C_3H_8) [118], but the high interaction energies require the use of high temperatures [71]. However, at higher temperatures the peaks are very much

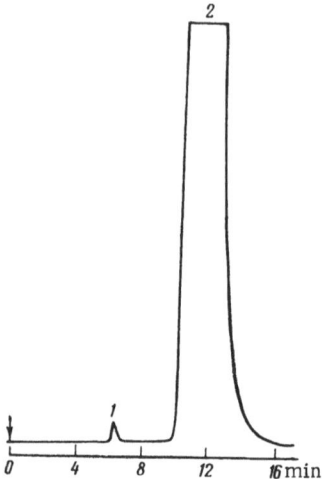

Fig. 104. Mixture of: 1) isobutane; 2) n-butane at 30°C on a 200 × 0.4 cm column of graphitized carbon black (flame-ionization detector).

broadened from inhomogeneity, including pores of various sizes (Fig. 101). Here adsorbed water has much less effect, because the adsorption is unspecific.

Activated charcoals may be used for the very important task of detection of traces of C_2H_2 in C_2H_4 in the manufacture of polyethylene [119] (Fig. 102). Here the acetylene peak emerges ahead of the main ethylene peak, which is very convenient for quantitative determination of trace amounts, since it is difficult to determine a minor component on the tail from a major one.

Microporous adsorbents (activated charcoals, silica gels, alumina gels) have large spreads in porosity, so unsymmetrical peaks with long tails are produced in the analysis of heavy hydrocarbons at low temperatures with large samples. Symmetrical peaks and shorter t_R are produced by modifying these adsorbents with small

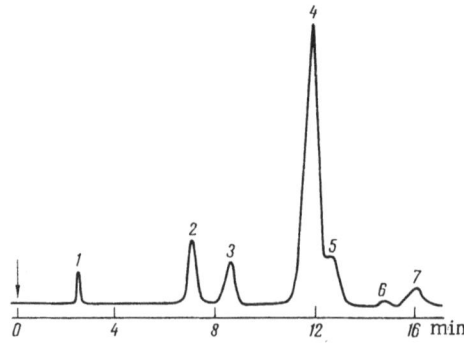

Fig. 105. Hydrocarbons at 30°C on a 200 × 0.4 cm column of graphitized carbon black (flame-ionization detector): 1) propylene; 2) isobutane; 3) α-butylene; 4) isobutylene; 5) cis-butylene; 6) divinyl; 7) trans-butylene.

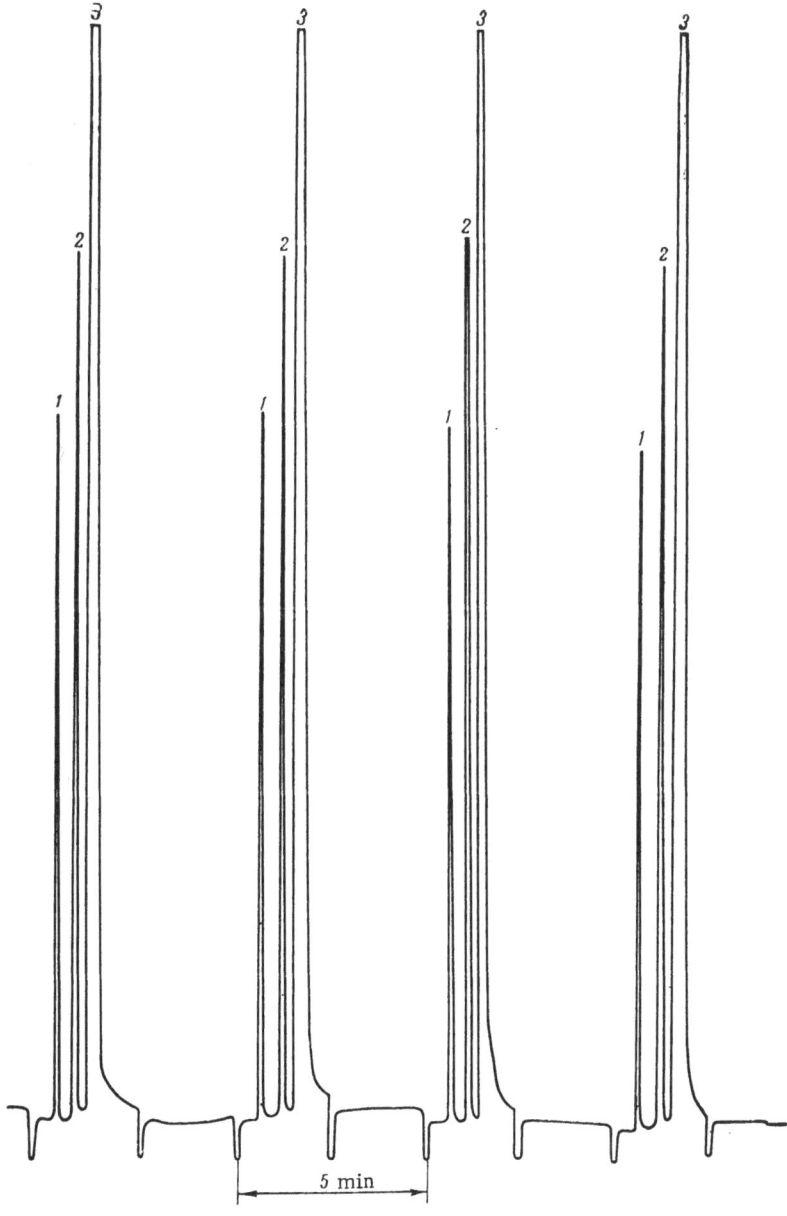

Fig. 106. Determination of traces of other gases in ethylene at 70°C on a 200 × 0.3 cm porous-glass column (0.4 ml sample, N₂ 60 ml/min, RKh-1 chromatograph): 1) methane (0.02%); 2) ethane (0.03%); 3) ethylene (99.5%).

amounts of liquids, which are virtually without effect on the separating capacity [120].

Lulova et al. [121] have used silica gel modified with glycerol plus soda in the analysis of hydrocarbon gases in the KhPA-2 chromatograph in relation to the production of polyethylene, especially in the analysis of light hydrocarbon condensate and also propane and propane—propene fractions.

Separation of isomeric butenes is more difficult, in view of the similarity in properties; it is usually difficult to separate but-1-ene and isobutylene on gas—liquid columns [122]. This mixture may be separated on macroporous silica gel modified with KOH, K_2CO_3, or K_2SiO_3 [123] (Fig. 103), though even better results are obtained with graphitized carbon black [124]. Here the uptake and V_R are governed mainly by the molecular geometry; isobutane is very readily separated from n-butane (Fig. 104) and but-1-ene from isobutylene (Fig. 105). The C_1-C_4 hydrocarbons have been separated with especially synthesized magnesium silicate [125] and also with silica gel [126]. Chemical modification of the silica gel

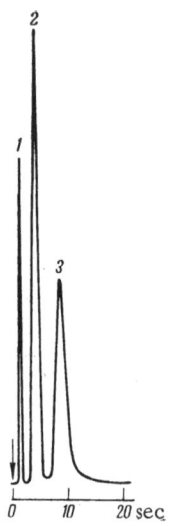

Fig. 107. Rapid analysis of hydrocarbons at 25°C on a 20 × 0.3 cm porous-glass column (H_2 40 ml/min, sample < 0.02 ml): 1) methane; 2) ethane; 3) ethylene.

(s of 200-300 m²/g) was produced by treatment with various alcohols in an autoclave at 200°C, which attached organic radicals to the surface (p. 39).

The best results were obtained by treatment with benzyl and lauryl alcohols, the hydrocarbons being separated at room temperature on columns 2 to 4 m long, which operated in a stable fashion. This method makes possible wide variation in the surface properties by choice of the alcohol.

The C_1-C_3 hydrocarbons have also been separated with porous glasses [127, 128], which are relatively easy to make more uniform than silica gel or alumina gel. Moreover, the porous structure and the depth of the porous layer are readily controlled via the chemical composition of the initial glass and the subsequent

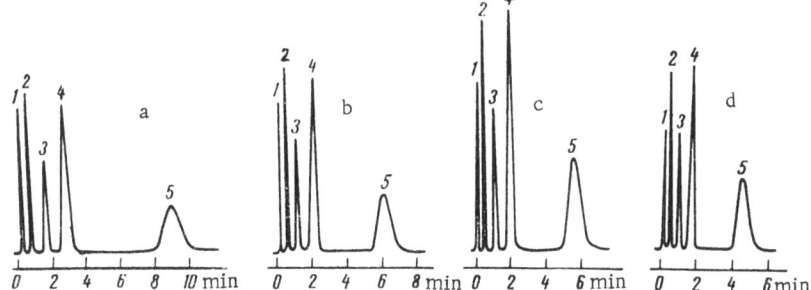

Fig. 108. Hydrocarbons on NaX zeolite at: a) 153°C; b) 165°C; c) 172°C; d) 180°C; column 50 × 0.45 cm, hydrogen 50 ml/min, 0.2 ml sample, katharometer [55]: 1) CH_4; 2) C_2H_6; 3) C_2H_4; 4) C_3H_8; 5) C_3H_6.

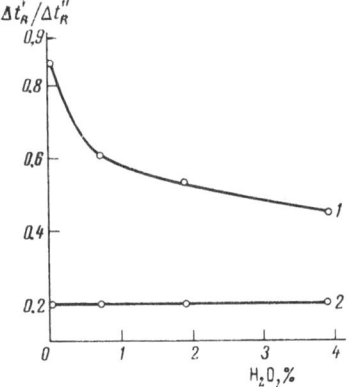

Fig. 109. Ratio of corrected retention times as a function of water adsorbed on NaX zeolite (150°C, 0.2 ml sample): 1) ethylene/propane; 2) ethane/propane.

treatment [129]. The high pore homogeneity provided by porous glasses means less peak broadening. The high selectivity and capacity of porous glasses provide especially favorable conditions for analysis for trace components in high-purity materials. The optimal ratio of adjacent components for chromatographic separation is one; ratios less than 1:1000 (i.e., trace components) impose severe demands on the separating capacity. Figure 106 shows a good separation of 0.02% CH_4 and 0.03% C_2H_6 in ethylene on porous glass. The high selectivity also means that rapid analyses can be done with short columns (Fig. 107).

Surface-porous glasses [131] have grains whose superficial porous layer is thin; they differ from other active adsorbents, which have through holes. These glasses substantially reduce the analysis time (Fig. 47), e.g., for C_1-C_3 hydrocarbons, without loss of resolution, which is of considerable importance in the use of gas chromatography in automatic process control. The performance of these columns is maintained up to linear speeds of 20 cm/sec.

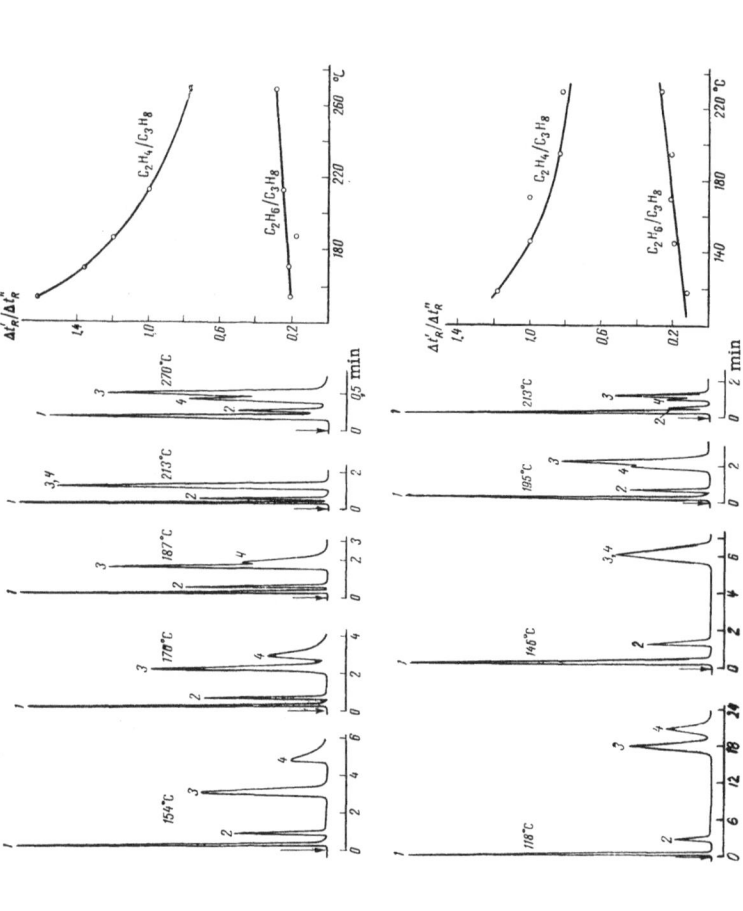

Fig. 110. Chromatograms of mixtures of: 1) methane; 2) ethane; 3) propane; 4) ethylene on: a) CaX; b) NaX at various temperatures; also ratio of corrected retention times for ethylene and ethane to that time for propane as a function of column temperature (50 × 0.4 cm column, 0.25-0.5 mm grain size, He, katharometer).

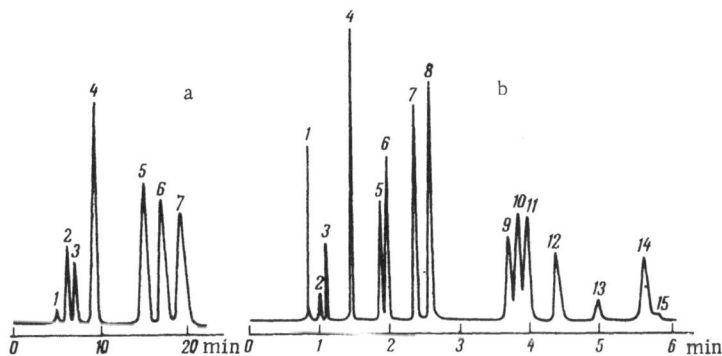

Fig. 111. a) Separation of: 1) methane; 2) ethane; 3) ethylene; 4) propane; 5) propylene; 6) isobutane; 7) n-butane at 25℃ on a micropacked alumina gel column 200 × 0.1 cm (carrier gas 4.65 ml/min, 0.25 ml sample, flow division in 1:700 ratio, flame-ionization detector); b) separation of: 1) methane; 2) ethane; 3) ethylene; 4) propane; 5) propylene; 6) acetylene; 7) isobutane; 8) n-butane; 9) but-1-ene; 10) trans-but-2-ene; 11) iso-butene; 12) cis-but-2-ene; 13) isopentane; 14) n-pentane; 15) butadiene (alumina gel column 10 m × 0.1 cm, 80℃, carrier gas 2.5 ml/min).

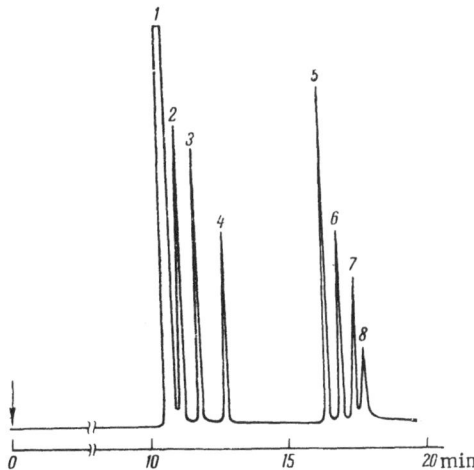

Fig. 112. Detection of minor components in me-thane on a glass capillary column 38 m × 0.22 mm at 0℃: 1) CH_4; 2) C_2H_6; 3) C_2H_4; 4) C_3H_8; 5) iso-C_4H_{10}; 6) C_3H_6; 7) n-C_4H_{10}; 8) C_2H_2.

Fig. 113. Mixture
of hydrocarbons at
25°C on a 3 m × 0.5
mm column, inter-
nal surface coated
with silica sol: 1)
cyclohexane; 2) 2,4-
dimethylpentane; 3)
benzene.

Hydrocarbon gases are analyzed on zeolites above 100°C [55, 132] (Fig. 108); the times of elution are very much dependent on the account of water previously adsorbed (Fig. 109). It has been found [56] that the order of emergence of ethylene and propane is reversed when the temperature of an anhydrous zeolite is raised from 120 to 190°C (Fig. 110).

The ratio of the corrected t_R for ethylene and propane increases with T for all zeolites, whereas that ratio for ethane and propane tends to remain constant or to increase slightly; this is due to the stronger T dependence of t_R for ethylene, on account of the specific interaction of the π-bond with the cations (Fig. 110).

Halasz and Heine [133] have used micropacked columns 2 to 10 m long filled with alumina gel of grain size 0.1-0.15 mm to separate the C_1-C_4 hydrocarbons; a 10 m column at 80°C separated 15 saturated and unsaturated hydrocarbons in 6 min, with good separation of all the isomeric butenes (Fig. 111). Higher linear gas speeds may be used with packed capillary columns; the minimal length equivalent to one theoretical plate is 0.3 mm for ethylene.

Svyatoshenko and Berezkin [134] used micropacked columns of high resolution with the light hydrocarbons; these approached capillary gas–liquid columns in performance, and their selectivity was almost that of ordinary packed columns [133, 134].

Gas adsorption may be used in capillary chromatography; glass capillaries with porous inner surfaces have been used [135] with hydrocarbons. This porous layer was produced on a special borosilicate glass by HCl etching. Bruner et al. [110, 136] produced similar glass capillaries by treatment with 20% NaOH, which completely separated traces of C_2-C_4 hydrocarbons present in methane (Fig. 112).

These porous capillary films have also been produced by deposition of hydrophilic silica sol [137], alumina [138], and fibrous

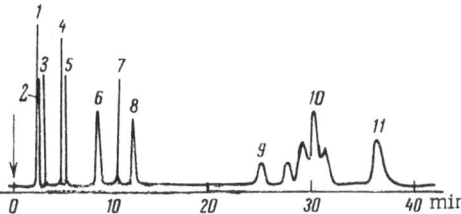

Fig. 114. Analysis of natural gas on an alumina capillary column at 100°C (15 m × 0.55 mm, carrier gas CO_2, 2 ml/min): 1) methane; 2) ethane; 3) propane; 4) isobutane; 5) n-butane; 6) neopentane; 7) isopentane; 8) n-pentane; 9) cyclopentane; 10) hexane isomers; 11) n-hexane.

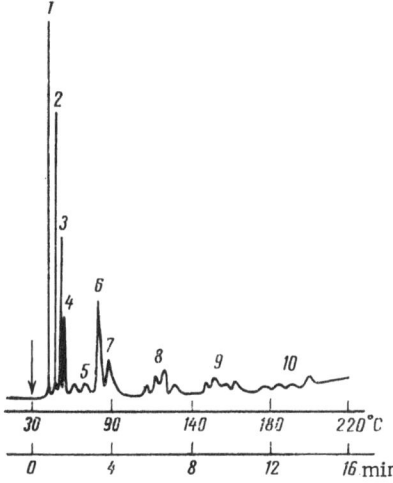

Fig. 115. Analysis of natural gas on a 6 m × 0.3 mm column containing glass spheres coated with fibrous boehmite and silica sol (spheres 60-80 mesh, He 25 ml/min, programmed temperature): 1) C_1; 2) C_2; 3) iso-C_4; 4) n-C_4; 5) neopentane; 6) iso-C_5; 7) n-C_5; 8) sum of C_6; 9) sum of C_7; 10) sum of C_8.

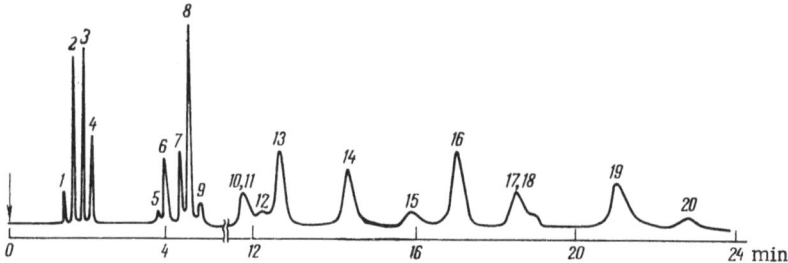

Fig. 116. C_1-C_4 hydrocarbons at 60°C on a 10 m × 0.25 mm column containing graphitized carbon black (~ 70 m²/g, grain size 0.15-0.20 mm) + 0.4% squalane: 1) methane; 2) acetylene; 3) ethylene; 4) ethane; 5) cyclopropane; 6) propylene; 7) propadiene; 8) propylene; 9) propane; 10) but-1-yne; 11) but-2-yne; 12) but-3-en-1-yne; 13) isobutane; 14) but-1-ene; 15) buta-1,2-diene; 16) isobut-1-ene; 17) n-butane; 18) cis-but-2-ene; 19) buta-1,3-diene; 20) trans-but-2-ene.

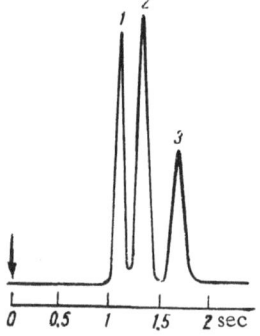

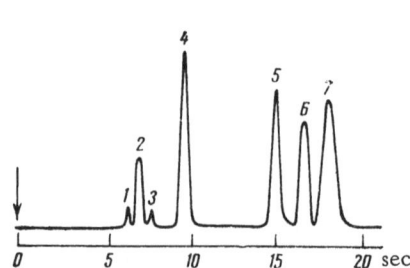

Fig. 117. Rapid analysis of a mixture of: 1) methane; 2) propane; 3) propylene at 25°C on a 20 cm × 0.4 mm column containing aerogel (1 μ particles, pressure differential 4 atm, linear gas speed 55 cm/sec, flame-ionization detector).

Fig. 118. Hydrocarbons at 25°C on a 160 cm × 0.4 mm column containing aerogel (1 μ particles, hydrogen 27 cm/sec, pressure differential 5 atm): 1) methane; 2) ethane; 3) ethylene; 4) propane; 5) propylene; 6) isobutane; 7) n-butane.

boehmite [139]; all gave good separation of hydrocarbon gases (Figs. 113-115). Packed capillary columns also give good separation; the C_1-C_4 hydrocarbons have been separated [140] with glass capillaries filled with graphitized carbon black (Fig. 116). Here the separation was in accordance with the number of carbon atoms and the molecular geometry.

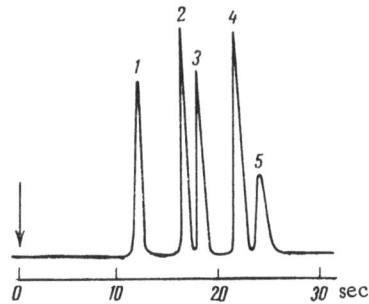

Fig. 119. The C_5 hydrocarbons at 71°C on a 160 cm × 0.4 cm column containing aerogel (1 μ particles, pressure differential 6 atm, linear gas speed 34 cm/sec): 1) n-pentane; 2) n-pent-1-ene; 3) 3-methylbut-1-ene; 4) 2-methylbut-1-ene; 5) 2-methylbut-2-ene.

Halasz and Gerlach [141] used a 0.4 cm capillary column only 20 cm long filled with finely divided aerogel (1 μ particles). The greatly reduced distance between grains diminishes the role of external diffusion, and so the performance is very good; here the minimum height corresponding to one theoretical plate was 0.12 mm. These columns are suitable for rapid analyses, though they do involve certain technical difficulties; for example, a mixture of methane, propane, and pro-

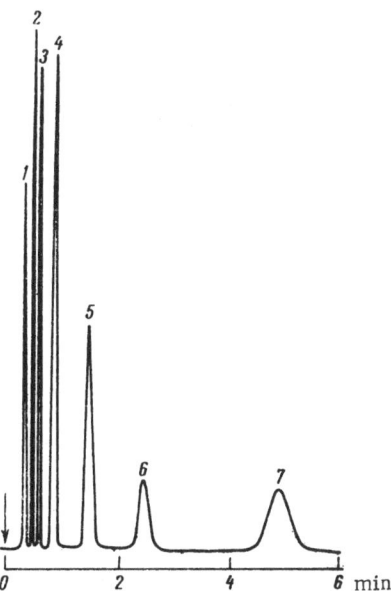

Fig. 120. Trace amounts (< 0.001%) of hydrocarbons in air at 70°C on a 100 × 0.4 cm column of activated alumina (flame-ionization detector, scale 10^{-11} A): 1) methane; 2) ethane; 3) ethylene; 4) propane; 5) propylene; 6) acetylene; 7) butane.

pene was separated in 1.7 sec (Fig. 117), with unadsorbed gas emerging in 0.7 sec. Again, such a column separated a mixture of C_1-C_4 hydrocarbons in 19 sec (Fig. 118). The separation of C_5 hydrocarbons was also very rapid (Fig. 119).

Gas-adsorption columns form the main way of determining trace amounts of hydrocarbon gases in air, since the minimum amounts detectable with gas-liquid columns are restricted by the volatility of the liquid [142]. Traces of C_1-C_4 in air and in liquid oxygen (0.08-0.2 ppm) have been measured with a β-ray ionization detector, and 0.004-0.07 ppm with a flame-ionization detector, on 10 ml samples [143]. Traces of C_1-C_6 hydrocarbons in air and hydrogen (0.01-100 ppm) have been determined on deactivated alumina gel with a flame-ionization detector with preliminary concentration [143]. Traces (0.01-0.1 ppm) of hydrocarbons in oxygen and air have similarly been determined [144] (Fig. 120).

Gas adsorption is widely used in the analysis of hydrocarbon gases in the production of polyethylene and polypropylene [145], in the determination of traces of hydrocarbons in air-liquefying plants (especially traces of acetylene in liquid oxygen [146]), in the analysis of fuel gases [147], in analysis of a methane—hydrogen fraction [148], in examination of dehydrogenation products from butane [149], exhaust gases [150], and dissolved hydrocarbons in liquid and porous bodies [151].

Gas adsorption chromatography is of particular importance in geochemical exploration, in the determination of traces in natural gases and in the products from cores [152].

ANALYSIS OF LIQUID MIXTURES

Gas-adsorption chromatography has as yet hardly been used for analysis of liquids, on account of the lack of a selection of nonporous and homogeneous macroporous adsorbents. The components of a liquid are retained for long times and are eluted as highly unsymmetrical peaks (even at high temperatures) if ordinary microporous adsorbents are used.

Chapter III deals with methods of making macroporous silica gels. Kiselev et al. [153-157] have shown that these can be used in gas-adsorption chromatography; various methods of chemical modification have been developed, such as alkali treatment [160] and attachment of alkylsilyl and alkoxy groups [158, 159].

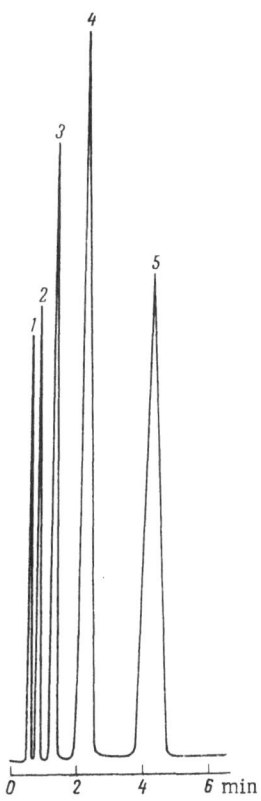

Fig. 121. Separation of n-alkanes at 100°C on a 100 X 0.4 cm column of macroporous silica gel (s ≈ 30 m^2/g, N_2 50 ml/min, flame-ionization detector): 1) pentane; 2) hexane; 3) heptane; 4) octane; 5) nonane.

Silica gel of mean pore size over 500 Å may be used in this way, in particular for hydrocarbons [156] (Fig. 121). Here the V_R are directly proportional to s, since pore narrowing (Fig. 41) here has no effect.

Zhdanov [161] has shown that porous glasses can be made with large uniform pores; here the important features are the composition and the heat treatment before leaching, which control the degrees of dispersion of the silicate and borosilicate components. The pores after leaching can be altered with alkali solutions under special conditions, which leads to partial or complete disruption of the microporous skeleton formed by finely divided silica gel within the larger pores related to the structure of the initial glass [161].

The heats of adsorption for macroporous glasses are similar to those for silica gel, which indicates that the two have similar geometry and chemically similar surfaces [162]. Zhdanov et al. [162] first used macroporous glasses to separate liquids with boiling points up to 200°C. Figure 122 shows curves for aromatic hydrocarbons and the C_6-C_{10} saturated hydrocarbons for a 100 cm X 4 mm column. The peaks are quite closely symmetrical. Figure 123 shows two chromatograms for the analysis of technical isopropylbenzene; in both cases only the ethylbenzene peak is of interest, as the heavier components may be flushed out by reversing the flow after this has emerged. One analysis takes 5-8 min. The baseline is stable.

The absence of background allows one to use adsorption columns with programmed temperature without shift in the baseline, a difficulty that occurs with gas—liquid columns (the small changes

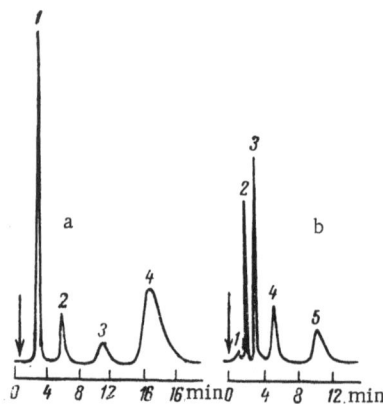

Fig. 122. Separation of mixtures on a 100 x 0.45·
cm column containing macroporous glass (sample
< 0.1 ml of vapor, carrier gas 60 ml/min): a)
120°C: 1) benzene; 2) toluene; 3) ethylbenzene;
4) isopropylbenzene; b) 102°C: 1) hexane; 2)
heptane; 3) octane; 4) nonane; 5) decane.

in gas flow speed associated with the temperature variation are not
recorded by ionization detectors). Figure 124 shows separation of
the normal C_1-C_9 hydrocarbons on macroporous silica gel in this
way (with no baseline shift), the normal gas scheme being used.

Figure 126 shows that t_R (from the maxima) is independent
of sample size for macroporous silica gel of uniform structure, so
the adsorption isotherm is of Henry type.

Macroporous unmodified silica gels of s from 20 to 100 m²/g
can be used in the analysis of compounds showing little or no specific interaction with the surface OH; such gels may be used with
saturated and aromatic hydrocarbons and similar compounds. Figure 127 illustrates this for aromatic hydrocarbons.

Silica gels may also be used with unsaturated hydrocarbons,
though this requires greater surface chemical purity, in order to
avoid catalytic reactions at inclusions containing Al and Fe. These
requirements are met by Aerosilogels [258, 259]; Fig. 125 shows
an example for aromatic hydrocarbons with program temperature
control.

Hydrophilic and hydrophobic silica sols can be deposited
within metal capillary columns [137, 163] to give capillary adsorp-

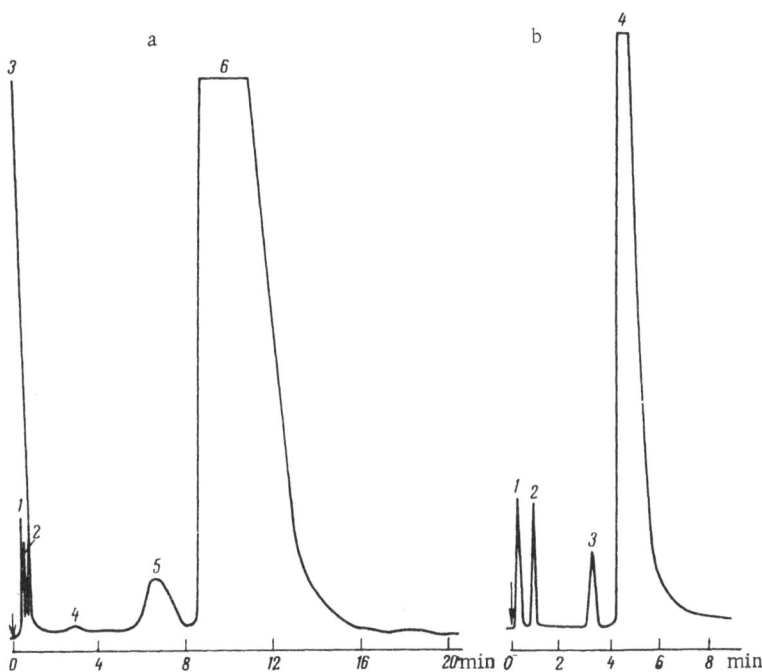

Fig. 123. Determination of minor components in technical isopropylbenzene on a 100 × 0.45 cm column containing macroporous glass (grain size 0.25-0.5 mm, H_2 80 ml/min, flame-ionization detector): a) 120°C: 1) propane; 2) propylene; 3) benzene; 4) toluene; 5) ethylbenzene; 6) isopropylbenzene; b) 150°C: 1) benzene; 2) toluene; 3) ethylbenzene; 4) isopropylbenzene.

tion columns, which are of high resolution for hydrocarbons. Used under the best conditions, these give separations comparable with those for capillaries bearing films of liquid. Figure 128 shows the separation of saturated and unsaturated C_1-C_6 hydrocarbons. The emergence of cyclohexane ahead of 2,4-dimethylpentane (cyclohexane has the higher boiling point) indicates that adsorption predominates in the separation (nonplanar structure is not favorable). A hydrophobic surface is indicated by the elution of benzene ahead of 2,4-dimethylpentane, since benzene is eluted much later on hydrophilic columns. A hydrophobic column of this type gives a good separation of the isomeric xylenes at 0°C (Fig. 129) and of a mixture of hydrocarbons with boiling points from 28 to 114°C. This separation with a liquid capillary column requires a length of at least 60 m.

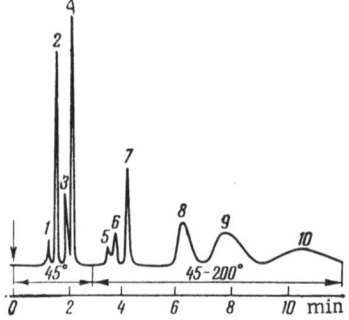

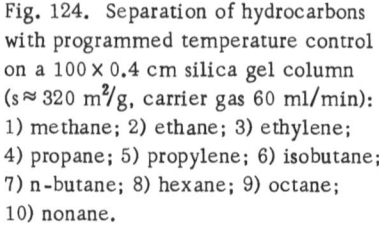

Fig. 124. Separation of hydrocarbons
with programmed temperature control
on a 100 X 0.4 cm silica gel column
($s \approx 320$ m^2/g, carrier gas 60 ml/min):
1) methane; 2) ethane; 3) ethylene;
4) propane; 5) propylene; 6) isobutane;
7) n-butane; 8) hexane; 9) octane;
10) nonane.

Fig. 125. Chromatogram on a 100 X 0.3
cm column of Aerosilogel ($s = 75$ m^2/g,
gas 50 ml/min, temperature 100-280°C,
heating rate 35°C/min) for: 1) benzene;
2) ethylbenzene; 3) decalin; 4) naphthalene; 5) diphenyl; 6) phenanthrene; 7)
terphenyl.

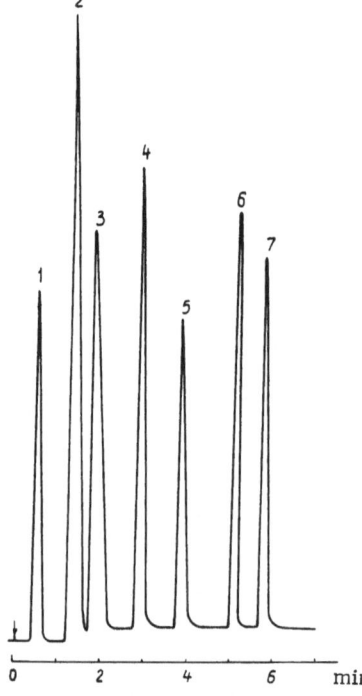

Fibrous boehmite, AlO(OH), forms platelets of thickness 50
Å and length ~1000 Å; it can also be used in separating liquid mixtures [164] when mounted on a support or on the inside of a capillary. Figure 130 shows the rapid separation of mixed Freons on
such a column, while Fig. 131 shows the detection of trace impurities in spectrally pure heptane. Bruner and Cartoni [136] obtained
an exceptionally high resolving power for glass capillaries with
porous inner surfaces modified with squalane; cyclohexane was
completely separated from deuterocyclohexane (Fig. 132).

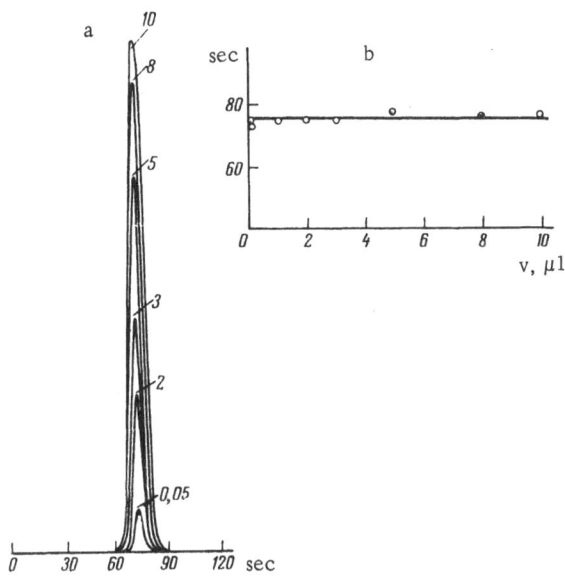

Fig. 126. a) Peaks produced by octane on macroporous silica gel at 100°C with different sample sizes; b) retention time as a function of sample size. The volumes of the samples in μl are given on the peaks.

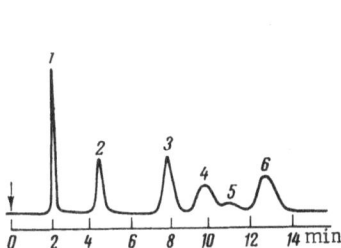

Fig. 127. Hydrocarbons at 120°C on a 200 × 0.4 cm column of macroporous silica gel (s ≈ 30 m²/g): 1) benzene; 2) toluene; 3) ethylbenzene; 4) m-xylene; 5) o-xylene; 6) isopropylbenzene.

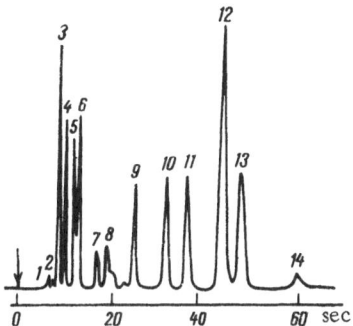

Fig. 128. Hydrocarbons at 27°C on a capillary adsorption columns: 1) methane; 2) ethane; 3) propane; 4) propylene; 5) isobutane; 6) n-butane; 7) but-1-ene; 8) isopentane; 9) n-pentane; 10) pent-1-ene; 11) 3-methylpent-1-ene; 12) 2-methylbut-1-ene; 13) 2-methylbut-2-ene; 14) n-hexane.

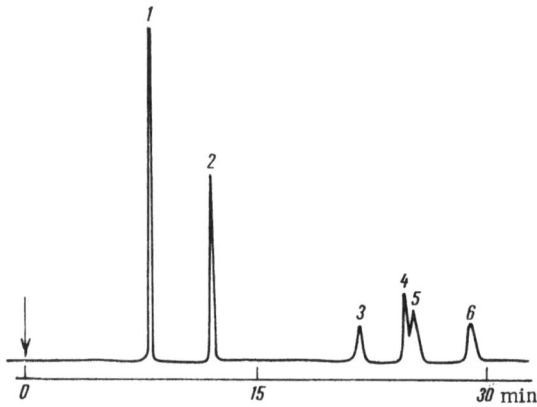

Fig. 129. Hydrocarbons at 20°C on a 60 m X 0.25 mm capillary coated with hydrophobic silica sol: 1) benzene; 2) toluene; 3) ethylbenzene; 4) m-xylene; 5) p-xylene; 6) o-xylene.

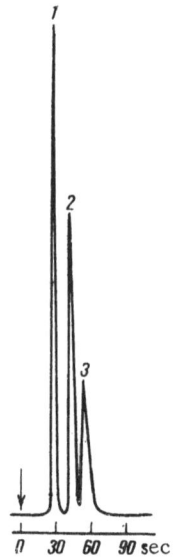

Fig. 130. Separation of Freons at 100°C on a 1 m column filled with C-22 support bearing boehmite (He 50 ml/min): 1) F-12; 2) F-144; 3) F-11.

It is inadvisable to use silica gel, alumina gel, and porous glasses to separate compounds of groups B, C, and D because the V_R are large on account of the strong specific interaction with the surface OH groups, and also because the chemical and geometrical nonuniformities give rise to considerable broadening of the trailing edge. In the separation of nitrogen heterocyclics (group B) on alumina gel [165], the adsorbed molecules set at an angle to the surface, with the N atom (bearing the nonbonding pair) attached to the surface; in this they differ from the aromatic hydrocarbons, which have the plane of the molecule parallel to the surface. The V_R are reduced [165] if ring substituents or the general ring geometry cause difficulty in orientation of the molecule relative to the surface.

Graphitized carbon black or chemically modified silica gel may be used

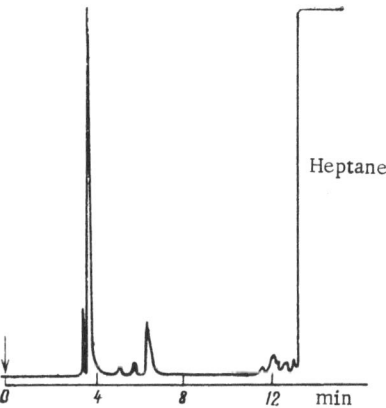

Fig. 131. Trace impurities in spectrally
pure heptane on a glass capillary column
90 m × 0.5 mm coated with fibrous boehmite and silica sol (He 12.5 ml/min, flame-
ionization detector).

with compounds showing strong specific interaction with OH
groups.

Kiselev [166-168] stressed the many uses of graphitized
carbon black, and the use of it has since greatly extended. Thermal blacks graphitized at about 3000°C are (Chapter,II) inert stable
nonspecific adsorbents with homogeneous surfaces (see literature
in Chapter II and also [166, 169]). The high surface homogeneity
of graphitized carbon blacks (especially thermal ones) and the nonspecific adsorption (p. 20 et seq.) provides fairly symmetrical peaks
(with small samples) for virtually all classes of compound. The
absence of internal porosity provides the best scope for rapid mass
transfer and hence a column of high performance.

Columns of graphitized carbon black have been used in research on the material [168, 170-174] and for analytical purposes.
The carbon black has [140, 175] been deposited as a thin film on a capillary column, which has been used in rapid analysis of aromatics and
compounds containing O and Cl (Fig. 133). Isomeric C_5-C_9 hydrocarbons have [140, 175] been separated with capillaries filled with
carbon black (Fig. 134). Pope [176] has used columns containing
carbon black deposited as a uniform layer on polyethylene spheres,
but the latter were not thermally stable. Halasz and Horwath [177]
have used columns with graphitized carbon black on glass spheres.

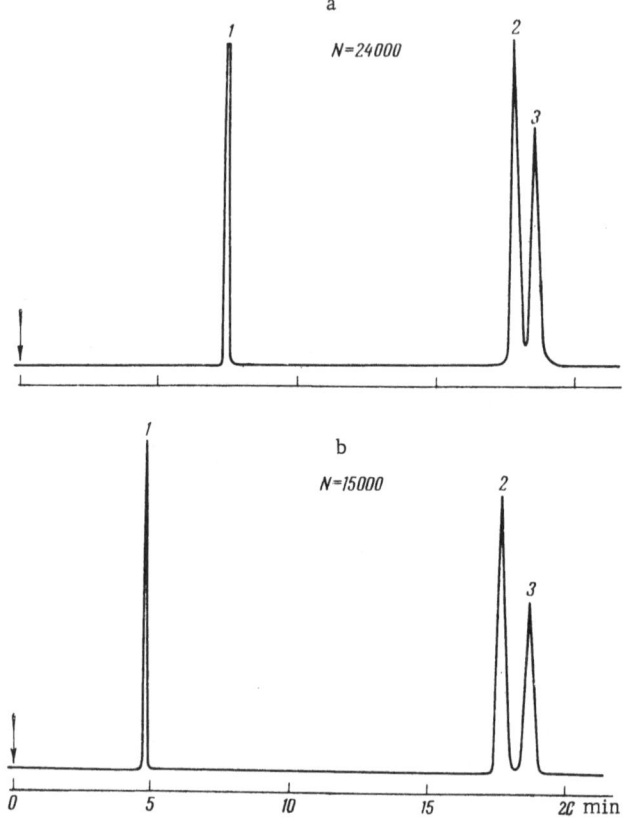

Fig. 132. Glass capillary column 38 m × 0.22 mm with a porous layer treated with squalane: a) 1%; b) 4% (N_2 0.8-1.25 ml/min) at 40°C separating: 1) methane; 2) deuterocyclohexane; 3) cyclohexane.

Graphitized carbon black has [171, 178] been mounted in the large pores of a support (macroporous silica gel with the surface modified by attachment of trimethylsilyl groups); but the carbon black may be used directly (without a support), since fairly stable aggregates are formed on manipulation [173], and the fraction of grain size 0.2-0.5 mm may be isolated

The absence of specific properties in graphitized carbon black (Chapter II, p. 20 et seq.) means that it retains more strongly the saturated hydrocarbons than polar or nonpolar substances of similar geometrical structure and molecular weight that can be attached to localized charges. Figure 4 (Chapter II) shows, for in-

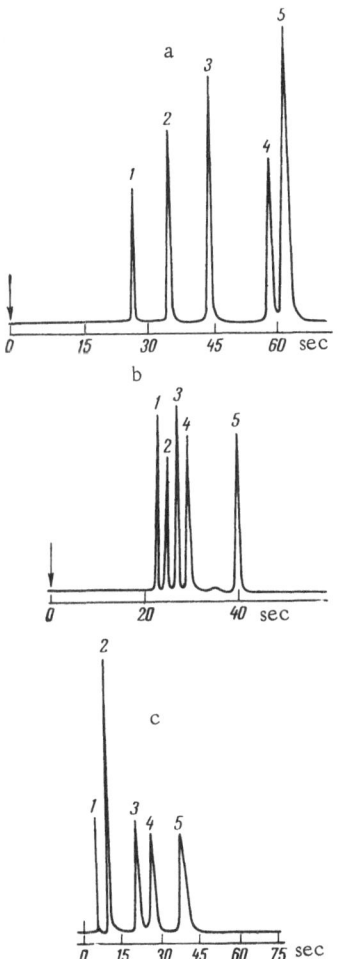

Fig. 133. Separations with a capillary column coated with graphitized carbon black (s≈70 m²/g, flame-ionization detector): a) column 8 m × 0.4 mm at 80°C: 1) methane; 2) dichloromethane; 3) trichloromethane; 4) CF₂Cl₂; b) column 15 m × 0.5 mm at 210°C, carrier gas 7.7 ml/min: 1) methanol; 2) propan-2-ol; 3) diethyl ether; 4) ethyl acetate; 5) n-hexane; c) column 15 m × 0.25 mm at 245°C, H₂ 2 ml/min: 1) benzene; 2) toluene; 3) ethylbenzene; 4) m-xylene; 5) o- and p-xylenes.

stance, that n-decane is retained more strongly than cyclohexanol, aniline, nitrobenzene, and acetophenone; while Fig. 135 shows that the normal hydrocarbons are held more strongly than aromatic ones with the same number of carbon atoms.

The material is of considerable interest for separating oxygen compounds in groups B and D (pp. 13-15 of Chapter II): water, alcohols, acids, aldehydes, ketones, ethers, esters, and other substances with the electron density localized at the periphery of functional groups (amines, pyridine, and other compounds containing N and S), because graphitized carbon black differs from glasses and macroporous silica gels in adsorbing these much more weakly (relative to the normal alkanes). Many compounds showing strong specific adsorption at OH groups can scarcely be eluted from columns containing silica gel or porous glass below 100°C, whereas on graphitized carbon black they emerge within 3 min at 30°C on a 100 cm × 0.5 mm column (Fig. 136).

Water and ammonia are only weakly bound by graphitized carbon black [173, 174]; alcohols and acids also have relatively small V_R [179, 180]. Figure 137 illustrates rapid

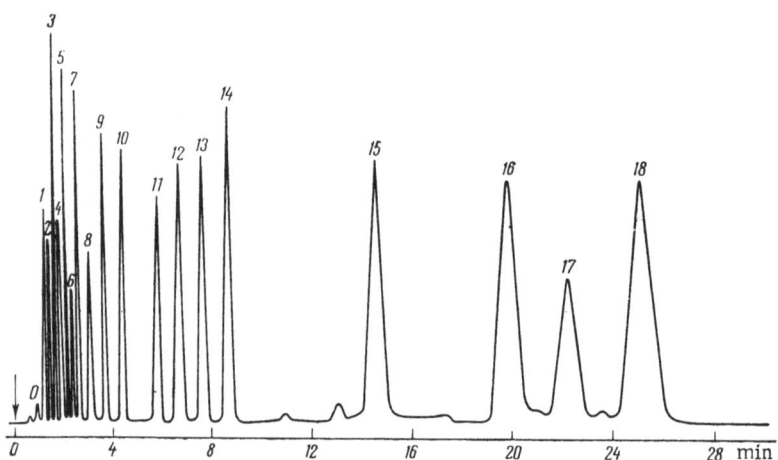

Fig. 134. Separation of C_5-C_9 hydrocarbons at 150°C on a 2 m × 0.25 mm glass capillary column filled with graphitized carbon black (0.15-0.20 mm) bearing 0.4% squalane: 0) methane; 1) cyclopentane; 2) isopentane; 3) n-pentane; 4) 2,2-dimethylbutane; 5) cyclohexene; 6) 4-methylpent-2-ene; 7) benzene; 8) 2,3-dimethylbuta-1,3-diene; 9) 2,3-dimethylbut-2-ene; 10) 2,4-dimethylpentane; 11) 2,4,4-trimethylpent-2-ene; 12) hept-2-ene; 13) 2,2,4-trimethylpentane; 14) n-heptane; 15) 2-ethylhept-1-ene; 16) oct-1-ene; 17) 2,6-dimethylhept-3-ene; 18) n-nonane.

separation of the C_1-C_4 alcohols at 118°C, while Fig. 138 illustrates separation of the normal C_5-C_{10} alcohols. The C_5-C_{10} fatty acids are separated in 3 min at 250°C (Fig. 139). Although the curves of Figs. 137-139 were recorded by katharometer, i.e., relatively large samples were used, the peaks are reasonably symmetrical. The fatty acids at these levels give somewhat broadened leading edges from interaction between acid molecules (Chapter IV, pp. 113-114). Even better results are obtained with ionization detectors.

High-boiling polar compounds such as glycols can [179, 180] be separated at 220-245°C in short times on graphitized carbon black (Fig. 140). Under these conditions, glycerol is eluted after diethylene glycol, the leading edge being broadened, not the trailing one.

The material has been used [179, 180] to separate amines (Fig. 141), a water–methanol–formaldehyde mixture (Fig. 142), aniline mixed with water and other compounds (Fig. 143), phenol mixed with cresols (Fig. 144), and mixtures of aromatic compounds (Fig. 145). Figure 146 shows that chlorinated compounds are eluted

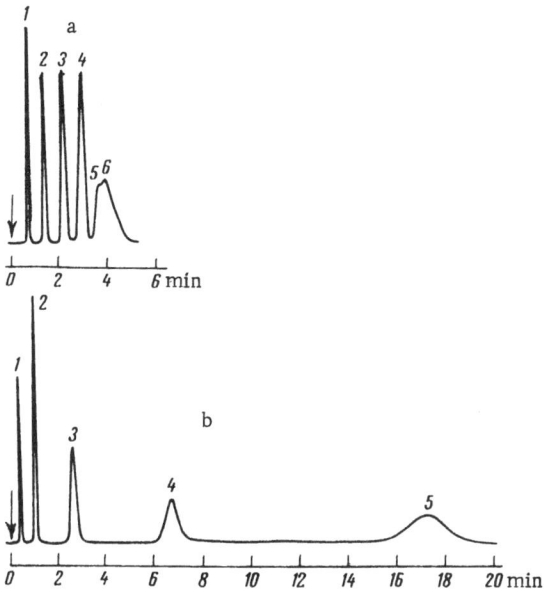

Fig. 135. Aromatic and normal hydrocarbons at 150°C on a 100 × 0.5 cm column of graphitized carbon black (s ≈ 7.6 m^2/g, He 60 ml/min, katharometer): 1) benzene; 2) toluene; 3) ethylbenzene; 4) isopropylbenzene; 5) and 6) xylenes; b): 1) hexane; 2) heptane; 3) octane; 4) nonane; 5) decane.

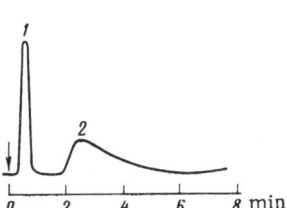

Fig. 136. Air (1) and water vapor at 30°C on a 100 × 0.5 cm column of graphitized carbon black (He 50 ml/min).

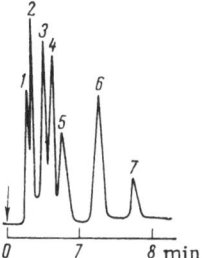

Fig. 137. Alcohols at 91°C on a 100 × 0.5 cm column of graphitized carbon black (He 50 ml/min): 1) methanol; 2) ethanol; 3) isopropanol; 4) n-propanol; 5) t-butanol; 6) i-butanol and s-butanol; 7) n-butanol.

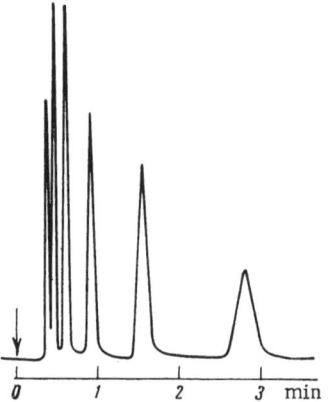

Fig. 138. Normal C_5-C_{10} alcohols at 250°C on a 100 × 0.5 cm column of graphitized carbon black (He 50 ml/min, katharometer).

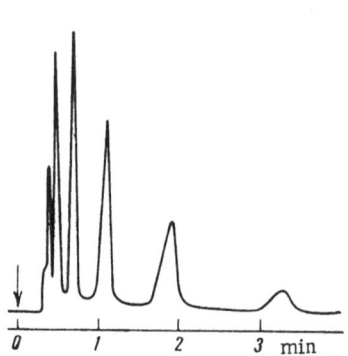

Fig. 139. Normal C_4-C_{10} fatty acids at 250°C on a 100 × 0.4 cm column of graphitized carbon black (He 50 ml/min).

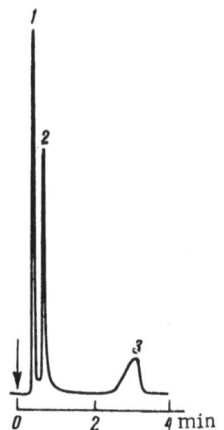

Fig. 140. Glycols at 220°C on a 200 × 0.5 cm column of graphitized carbon black (katharometer detector): 1) ethylene glycol; 2) diethylene glycol; 3) triethylene glycol.

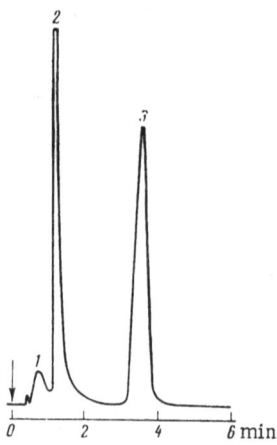

Fig. 141. Column at 110°C and 200 × 0.5 cm of graphitized carbon black (katharometer detector): 1) water; 2) diethylamine; 3) triethylamine.

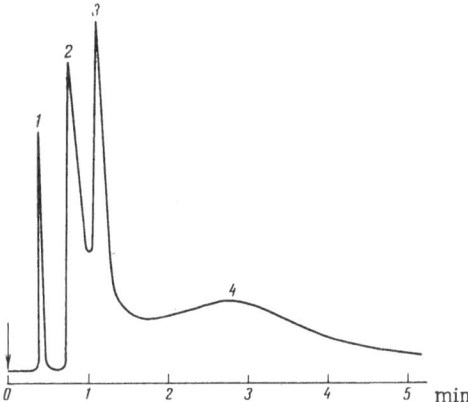

Fig. 142. Column 70 x 0.3 cm of graphitized carbon black (s ≈70 m²/g, He 60 ml/min, katharometer) at 40°C: 1) air; 2) water; 3) methanol; 4) formaldehyde.

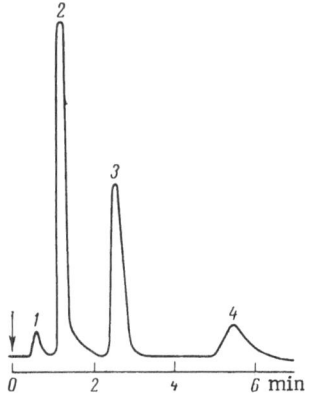

Fig. 143. Column at 180°C and 200 x 0.5 cm of graphitized carbon black (katharometer detector): 1) water; 2) aniline; 3) methylaniline; 4) dimethylaniline.

in accordance with the number and disposition of the chlorine atoms, not with the boiling points. It has been pointed out (p. 28 et seq., Chapter II) that dispersion interaction is here responsible for the adsorption, and this is dependent on the spatial disposition of the atoms, so geometry and orientation dominate gas chromatography on graphitized carbon black [180, 181], V_R being determined mainly by the total polarizability, the number of electrons, and the orientation of the links, with the boiling point, molecular weight, and so on much less important [179-181].

Isopropylbenzene is eluted before the xylenes, whereas n-propylbenzene is eluted later (Fig. 147) [181], which is due to features of the geometrical structure. The isomeric butylbenzenes are also well separated, on account of the difference in geometry

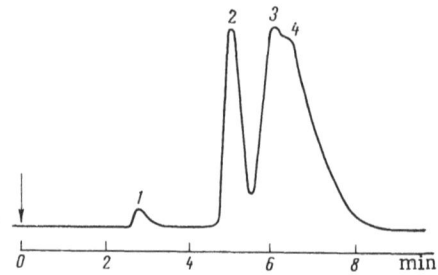

Fig. 144. Column 200 x 0.4 cm of graphitized
carbon black (katharometer detector) at 190°C:
1) phenol; 2) o-cresol; 3) m-cresol; 4) p-cresol.

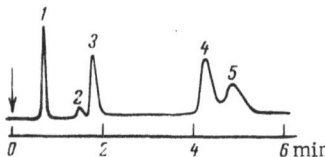

Fig. 145. Column at 240° and 200
x 0.5 cm of graphitized carbon black
(katharometer detector): 1) benzene;
2) and 3) cis- and trans-decalin; 4)
quinoline; 5) naphthalene (He 50
ml/min).

[180, 181]. For the same reason, thiophen is eluted before
benzene, although it exceeds
benzene in boiling point and molecular weight; this allows the
detection of 0.5% thiophen in
benzene (Fig. 19).

The m-, p-, and o-xyl-
enes emerge from graphitized
carbon black in an order the
reverse of that given by gas–
liquid columns with polar phases
[182] and adsorption columns with specific adsorbents [183].

Table 10 shows that m-xylene can present only one carbon
atom to the basal plane of graphite, whereas o- and p-xylene can
present two; this occurs because the van der Waals radius of a CH_3
group (2.0 Å) is greater than half the thickness of a benzene ring
(1.85 Å). Hence m-xylene runs considerably ahead of the other xylenes on graphitized carbon. The o- and p-xylenes can present equal
numbers of carbon atoms to the graphite, so the difference in adsorption energies is small and the separation is poor, at least for
the short column with material of low s used in [181]; but a column
equivalent to only 2000 theoretical plates gave complete separation
of all three, whereas a capillary column bearing squalane requires
to be equivalent to at least 70,000 theoretical plates to do this.
Again, p-xylene is readily separated from o-xylene on type II (specific) adsorbents, on account of the considerable difference in the

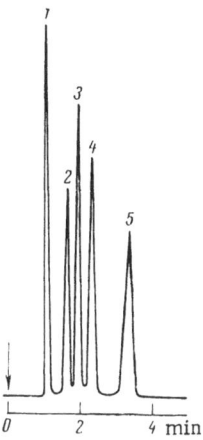

Fig. 146. Column at 95°C
and 200 × 0.5 cm of graphitized carbon black (ka-
tharometer detector): 1)
methylene chloride; 2)
propyl chloride; 3) chloro-
form; 4) dichloroethane;
5) CCl$_4$ (He 50 ml/min).

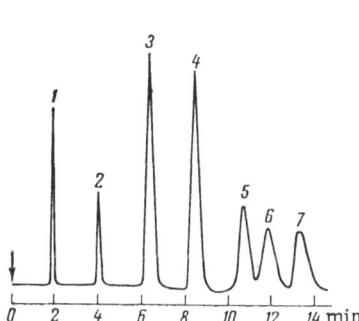

Fig. 147. Column at 150°C and 200
× 0.5 cm of graphitized carbon black
(katharometer detector): 1) benzene;
2) toluene; 3) ethylbenzene; 4) isopropylbenzene; 5) m-xylene; 6) o-
and p-xylenes; 7) n-propylbenzene
(N$_2$ 50 ml/min).

electron-density distributions. Composite columns
containing adsorbents of
types I and II are very useful in such cases.

Compounds such as
ferrocene are bound only
relatively weakly by graphitized carbon black, be-
cause the sandwich struc-
ture allows relatively few
atoms to contact the sur-
face (Fig. 148).

Graphitized carbon
black has also been used
[184] to separate terpenes.

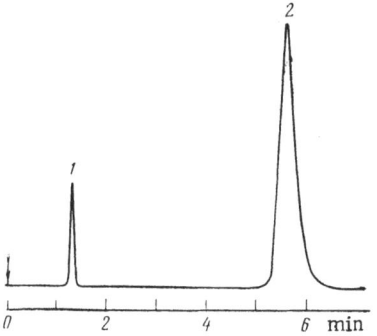

Fig. 148. Column 200 × 0.2 cm at 230°C
of graphitized carbon black (grain size
0.25-0.5 mm, flame-ionization detector): 1) ferrocene; 2) n-decane.

Figure 149 shows that the energy of interaction of a terpene here
decreases as the molecular geometry becomes more complicated

Table 10. Molecular Structure of Xylenes and Possible
Orientations on the Basal Plane of Graphite

Isomer	Structural formula*	No. of ring C atoms possibly in contact	No. of ring C atoms without possibility of contact
m-Xylene		1	5
o-Xylene		2	4
p-Xylene	or	2	4

*Crosses indicate positions that can contact the surface.

Table 11. Relative V_R for Terpenes on Graphitized
Carbon Black and in Various Liquid Phases*

Terpene	Fixed Phase					
	Graphitized carbon black Sterling MT, 3000°	20M Poly-ethylene glycol	Tricresyl phosphate on Sterhamol	Carbowax 4000	DS-550 silicone on celite	Dodecyl phthalate
	$l\dagger = 1.2$ m 150° C	$l = 1.8$ m 65° C	$l = 5.0$ m 100° C	$l = 2.0$ m 130° C	$l = 3.0$ m 156° C	$l = 2.0$ m 150° C
Camphene	0.95	1.4	1.24	1.28	1.13	1.21
α-Pinene	1.00	1.0	1.00	1.00	1.00	1.00
Camphor	1.52	—	—	—	—	—
Borneol	1.90	—	—	—	—	7.10
Fenchone	2.75	—	—	—	—	3.77
Fenchol	2.79	—	—	—	—	5.13
Δ^3-Carene	3.92	—	1.97	1.82	1.90	1.66
Sabinene	8.70	2.06	—	—	—	—
Limonene	8.70	3.33	2.48	2.29	1.95	1.90
p-Cymene	8.70	6.0	3.40	3.15	2.42	2.13
α-Terpineol	10.7	—	—	—	—	7.98
Menthone	13.3	—	—	—	—	—
Menthol	16.6	—	—	—	—	—
Carvone	21.3	—	—	—	—	—

*Data on liquid phase from [185].

†l = column length

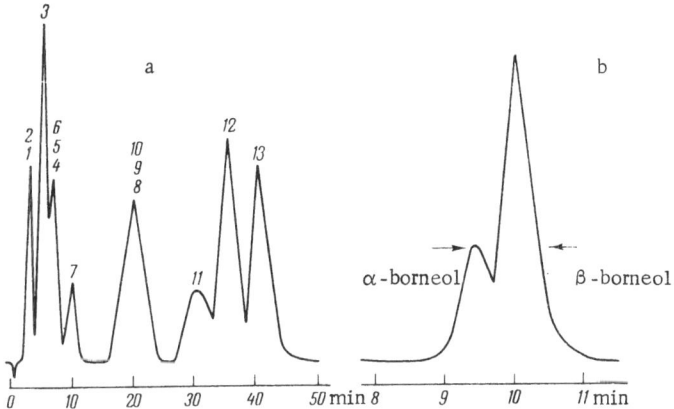

Fig. 149. Separation of a) terpenes: 1) carene; 2) α-pinene; 3) camphor; 4,5,6) borneol, fenchol, fenchone; 7) Δ^3-carene; 8,9,10) sabinene, limonene, p-cymene; 11) menthone; 12) menthol; 13) carvone; at 180°C; and b) α- and β-borneols at 153°C (1.2 m × 4 mm column of graphitized carbon black, carrier gas 43 ml/min).

(greater deviation from coplanarity). There are major difficulties in separating the terpenes because of the similarities in chemical structure and physical properties. Differences in boiling point are of great importance in gas–liquid chromatography the terpenes cause difficulty because many of them have similar boiling points. However, the molecules differ considerably in geometry, so they can be separated on this basis. The bicyclic terpene hydrocarbons are bound the least strongly, the size of the rings affecting V_R (if the second ring is formed in the 2,4-position of the six-membered ring, the compound is bound more firmly than that for the 1,4-position). The monocylcic terpene hydrocarbons are bound more strongly than the bicyclics, while the aliphatic terpenes are bound the most strongly. Table 11 compares the relative V_R for graphitized carbon black and for various liquid phases used for this purpose (see [185] for literature). Adsorption on graphitized black gives a much greater range in these quantities than does gas–liquid chromatography. Figure 149 shows spearations for some terpenes. Graphitized carbon black gives a poor separation when the molecules are similar and similarly oriented on the basal plane; in such cases the separation may be improved by combination with a specific adsorbent, e.g.,

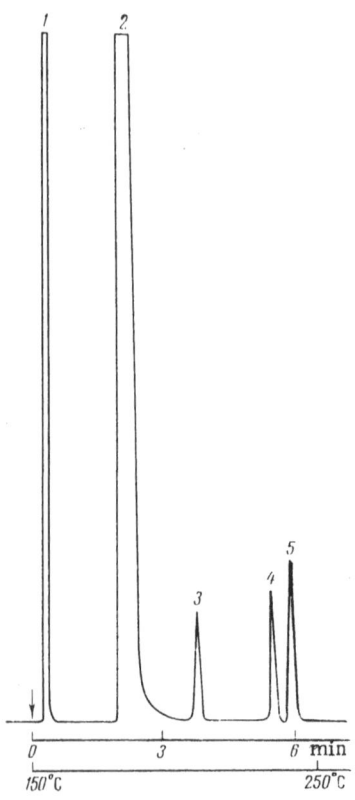

Fig. 150. Impurities in phenol on a 200 x 0.2 column of graphitized carbon black (s = 7.6 m^2/g, grain size 0.25-0.5 mm, programmed temperature variation from 150 to 250°C, rate 17 deg/min): 1) acetone (solvent); 2) phenol; 3) α-methylstyrene; 4) acetophenone; 5) isopropylbenzene.

macroporous silica gel with an appropriately hydroxylated surface.

Graphitized carbon gives no background, so it can be used with program temperature control up to very high temperatures; no differential gas circuit is needed [186]. Such control is especially useful if some components have large t_R.

Programmed temperature rise produces narrower peaks and higher peak concentrations; Fig. 150 shows the detection of minor components in phenol in this way, where all the peaks are relatively narrow.

Graphitized carbon black has been mounted on macroporous Teflon [186] to give more rapid mass transfer. The adsorption on the Teflon is much weaker than that of the black, so the Teflon can be considered as a nearly inert support. The analysis times for such columns (Fig. 151) are much less than those for columns of granulated graphitized black (compare Fig. 135); program temperature control may also be used, as with the $C_1 - C_{10}$ normal alcohols (Fig. 152).

Capillary graphitized-carbon columns can separate deuterated compounds. It has been shown [260] that a 9 m glass capillary column with graphitized carbon on the inner surface, and with the addition of a little squalane, will separate deuterocyclohexane from cyclohexane and deuterobenzene from benzene (Fig. 153). It has

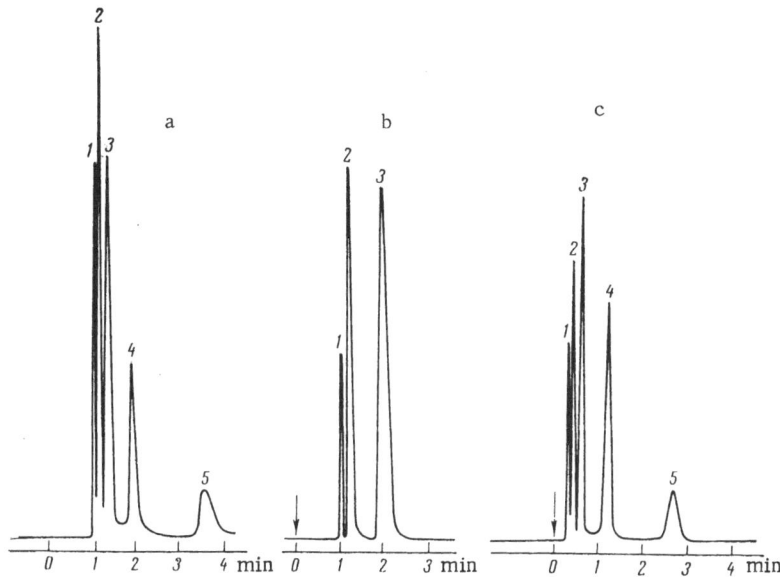

Fig. 151. Column 200 × 0.4 cm of graphitized carbon black on teflon at 160°C (gas nitrogen, flame-ionization detector): a) C_5-C_9 normal paraffins; b) mixture of: 1) benzene; 2) toluene; 3) p-xylene; c) normal C_4-C_8 alcohols.

also been shown [261] that pure graphitized thermal carbon black provides extremely high separating power for deuterated benzene, toluene, acetone, and pyridine (Fig. 154).

The separation temperature can be reduced considerably, and the selectivity can be adjusted, if graphitized carbon black is coated with a monolayer of an organic compound (Fig. 32) or with organic crystals, e.g., metal phthalocyanins.

Adsorbents other than graphitized carbon black are also widely used in gas chromatography. In particular, very interesting results have been obtained with adsorbents prepared by attaching organic cations by ion exchange to the surface of montmorillonite (bentone-34, dimethyldioctadecyl ammonium bentonite). The use of this in gas chromatography was described in 1957 by White [187], who considered that the separation of aromatic hydrocarbons on this is due to differences in adsorption. White and Cohen [188] called it the ideal adsorbent for gas chromatography. Hughes et al. [189] used bentone-34 to separate xylenes, cresols, and toluidines, though the peaks were somewhat unsymmetrical. This has subse-

Table 12. Performance in Separation of m- and p-Xylenes
on Columns with Various Fixed Phases

Column type	Fixed phase	Temp., °C	V_{R_1}/V_{R_2}	No. of theor. plates
Open capillary	Squalane	78	1.015	100,000
	Benzyldiphenyl	78	1.032	40,000
	Polyethylene glycol	50	1.035	35,000
	7,8-Benzoquinoline	78	1.07	8,500
Ordinary filled	Bentone-34 with silicon oil on diatomite support	70	1.265	800
	Graphitized carbon black	150	1.143	2,000

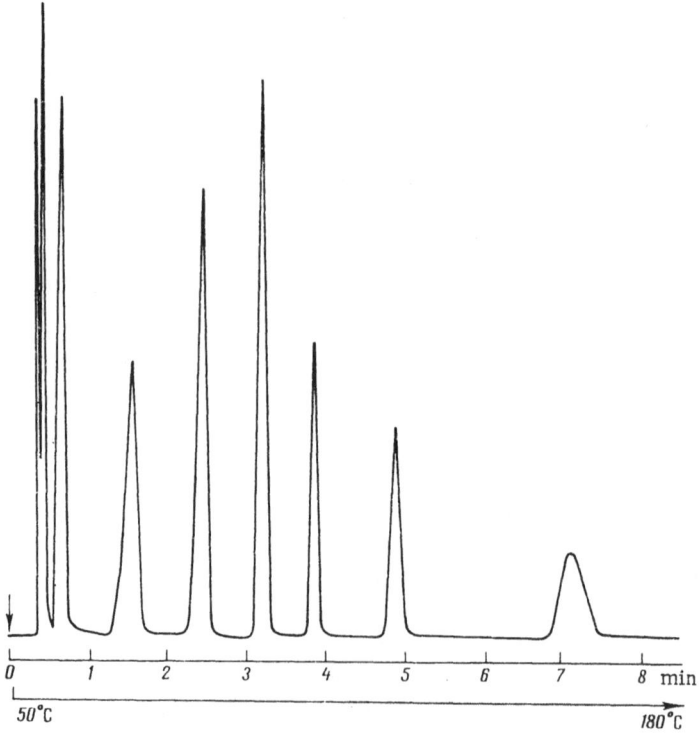

Fig. 152. Column 100 × 0.4 cm of graphitized carbon black on Teflon with programmed temperature (50-180°C) with normal C_1–C_9 alcohols (gas carrier nitrogen, flame-ionization detector).

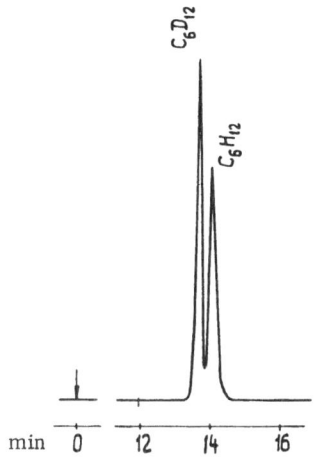

Fig. 153. Separation of deuterocyclohexane from cyclohexane
on a 9 m × 0.3 cm column coated
with graphitized thermal carbon
black (s = 7.6 m^2/g) containing
0.5% squalane at 25°C.

quently been overcome by modification with organic liquids, especially silicone oil [190] and phthalates [191] (Fig. 155). Bentone-34
has also been used to separated
chlorobenzenes [192] (Fig. 156),
methylnaphthalenes [193], and
polyphenyls [194] (Fig. 157).
An adsorbent based on bentone-34
has recently been proposed [195]
for the separation of isopropyl-
benzenc from ethylbenzene
(Fig. 158).

Bentone-34 is highly selective in separating isomers because
it operates mainly by adsorption
on the modifying layer, although

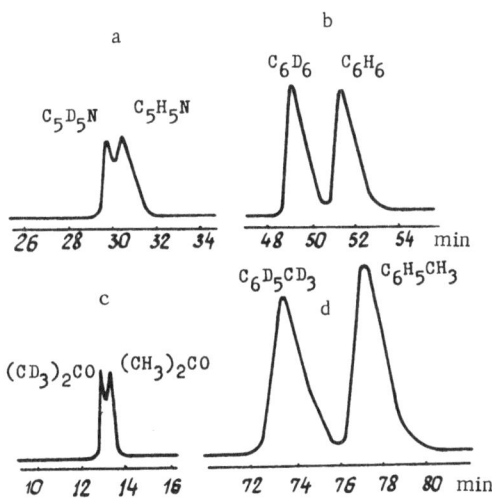

Fig. 154. Separation of deuterated compounds from
the corresponding undeuterated ones on a 9.6 m × 0.15
mm column coated with graphitized thermal carbon
black at: a) 87.5°C; b) 69.5°C; c) 87.5°C; d) 98.4°C.

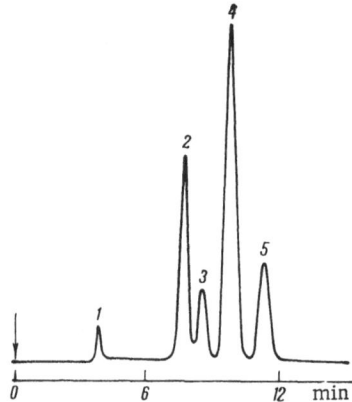

Fig. 155. Aromatic hydrocarbons at 75°C on a 180 × 0.5 cm column with 5% bentone-34 and 5% diisodecyl phthalate (He 100 ml/min): 1) toluene; 2) ethylbenzene; 3) p-xylene; 4) m-xylene; 5) o-xylene.

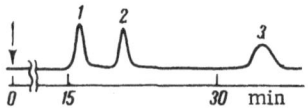

Fig. 156. Dichlorobenzenes at 150°C on a 360 cm column of bentone-34 + silicone oil (1:1): 1) p-dichlorobenzene; 2) m-dichlorobenzene; 3) o-dichlorobenzene.

the surface bears fairly long chains, between which the compounds could penetrate ("dissolve"). It has been shown (p. 28) that isomers, especially purely steric ones, are adsorbed differently on the planar surface of the adsorbent, in spite of their similarity in chemical and physical properties, because their links lie at different distances from the surface. Bentone consists basically of platy attachment of organic bases, and adsorption occurs on a fairly uniform surface.

The high selectivity of bentone-34 provides separation of m- and p-xylenes (usually difficult to separate) on a short column. Table 12 shows the column performance required to separate these xylenes with various fixed phases [192].

Favre and Kallenbach [196] have used various inorganic adsorbents in gas chromatography, including 44 inorganic salts and hydroxides. The separation performance was tested with a mixture of isomeric terphenyls. The salts included antimonates, borates, carbonates, nitrates, and phosphates. Some of these gave high resolution with relatively short t_R; the best results were obtained with LiCl and K_2CO_3. Such columns can operate at or above 500°C.

It is difficult to use some of these compounds because they have very small s; but dehydration of hydrates (or other coordination compounds) can give relatively high s. Rogers and Altenau [197, 198] used $Cu(Py)_2(NO_3)_2$, $Cu(NH_3)_4(NO_3)_2$, $Cu(Py)SO_4$, $CdSO_4$, $CuSO_4 \cdot H_2O$ (see Chapter VI on the details of the production of such adsorbents). A 1.7 m column of $Cu(Py)_2(NO_3)_2$ at 38°C separated

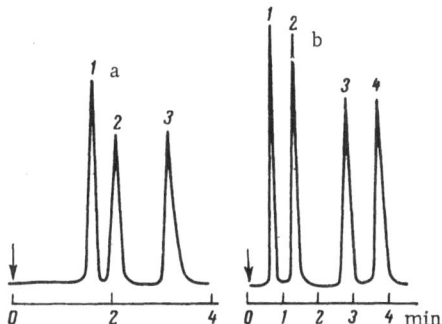

Fig. 159. Column 168 cm long of Cu(Py)$_2$(NO$_3$)$_2$
at 38°C: a) alcohols: 1) t-amyl; 2) isobutyl; 3)
n-butyl; b) ketones: 1) acetone; 2) methyl ethyl
ketone; 3) pentan-2-one; 4) pentan-3-one.

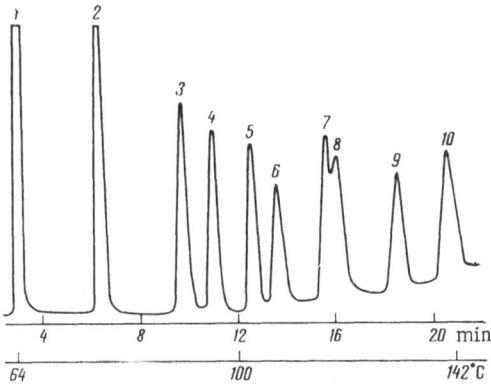

Fig. 160. Aromatic compounds on a 183 cm column
of Cu(Phen)SO$_4$ · H$_2$O with program temperature variation: 1) benzene; 2) toluene; 3) ethylbenzene; 4)
chlorobenzene; 5) o-xylene; 6) mesitylene; 7) bromo-
benzene; 8) cumene; 9) p-butyltoluene; 10) iodobenzene.

columns were ones of CdSO$_4$ and Cu(Py)$_2$(NO$_3$)$_2$; heats of adsorption
were measured for these, and large Q were found for compounds of
large t_R. The highest Q were for acetone, methyl ethyl ketone,
ethyl acetate, and ethyl propionate on Cu(NH$_3$)$_4$(NO$_3$)$_2$ and Cu(Py)SO$_4$,
being about 15 kcal/mole.

The Q for these adsorbents for aromatic and oxygen compounds show that the metal ion interacts with the ring π-electrons
or with the unpaired electrons on the oxygen atom.

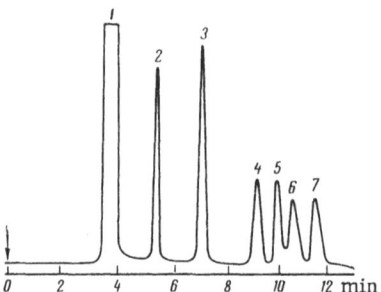

Fig. 161. Aromatic compounds on a column of diatomite bearing an anisotropic melt of p,p'-azoxyphenetol: 1) pentane (solvent); 2) benzene; 3) toluene; 4) ethylbenzene; 5) m-xylene; 6) p-xylene; 7) o-xylene.

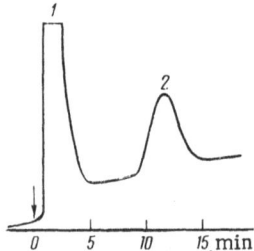

Fig. 162. Separation of: 1) TiCl$_4$ and 2) SbCl$_3$ at 240°C on a column containing a eutectic mixture of BiCl$_3$ and PbCl$_2$ mounted on diatomite.

The main disadvantage of these adsorbents is the low thermal stability. Altenau and Rogers [199] described the use of Cu(Phen)$_2$(NO$_3$)$_2$, Cu(Phen)Cl$_2$, Cu(Phen)SO$_4$ · H$_2$O, and Cu(Bipy) · (NO$_3$)$_2$, which contain 1,10-phenanthroline (Phen) and 2,2'-bipyridyl (Bipy); these do not decompose below 500°C, and there is virtually no increase in the background as the temperature is raised, so program temperature control can be used. Figure 160 illustrates the separation of aromatic hydrocarbons having a wide range of boiling points in this way with Cu(Phen) · SO$_4$ · H$_2$O.

Liquid crystals form stationary phases of considerable interest [200-202]. At 117°C, p,p'-azoxyanisole melts to give a liquid crystal, which at 135°C becomes an isotropic liquid. Partition coefficients and V_R have been measured for aromatic hydrocarbons for both states; the V_R differ in absolute magnitude and in temperature dependence.

In the solid state (room temperature up to the first transition point) the material gives good separation of cyclohexane, benzene, toluene, xylene, and the C$_5$–C$_{12}$ normal paraffins (m- and p-xylenes are almost unseparated). At 117°C the V_R of all these (relative to the weight of the p,p'-azoxyanisole) increase by over an order of magnitude.

There is a difference of 2–3 kcal/mole between the partial molar enthalpies of solution at infinite dilution for the anisotropic

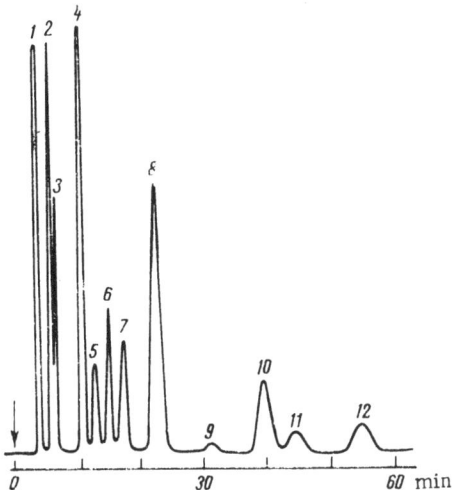

Fig. 163. Separation at 175°C on a 90 × 0.3
cm column of NaI on alumina: 1) trans-4-
methylpent-2-ene; 2) 1,5-hexadiene; 3) 2,2,4-
trimethylpentane; 4) benzene; 5) trans-oct-2-
ene; 6) cis-oct-2-ene; 7) n-nonane; 8) toluene;
9) n-decane; 10) ethylbenzene; 11) o-chloro-
toluene; 12) m-chlorotoluene.

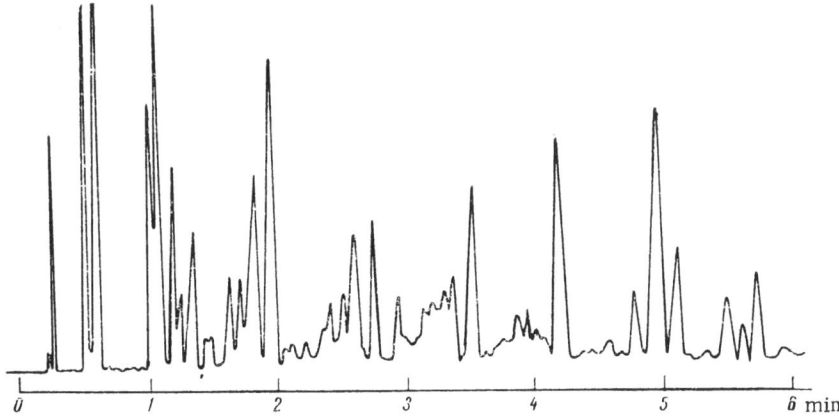

Fig. 164. Kerosene examined with program temperature variation (50 to 250°C) on
a 150 × 0.3 cm column with NaI on alumina.

and isotropic melts. The order of elution of m- and p-xyelene is
reversed at the second transition point.

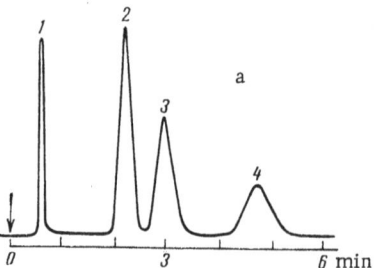

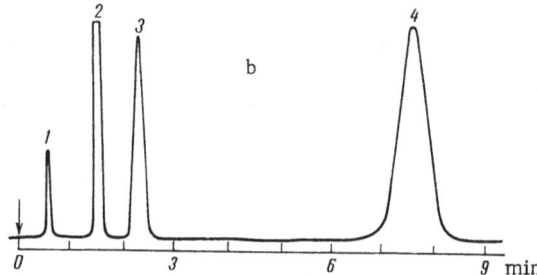

Fig. 165. Hydrocarbons at 100°C on: a) 45 cm × 3 mm
column of NaCl on Al$_2$O$_3$; b) 80 cm × 0.3 mm column
of NaI on Al$_2$O$_3$: 1) cyclohexane; 2) n-heptane; 3) iso-
octane; 4) benzene.

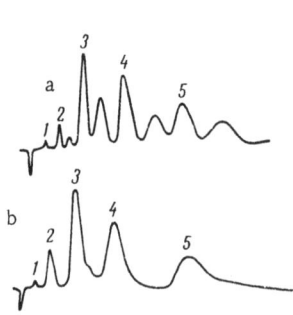

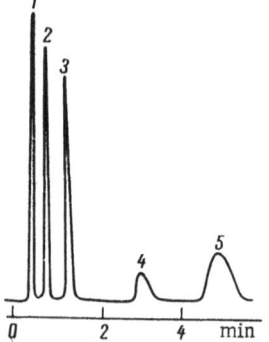

Fig. 166. C$_{10}$–C$_{14}$ normal alkanes
(1-5 respectively) and alkenes (not
numbered) on Al$_2$O$_3$ modified with:
a) NaOH; b) NaOH+CuCl$_2$ (in the
latter cases the alkenes do not
emerge from the column).

Fig. 167. Porous-polymer
column 180 × 0.5 cm at 158°C
(He 60 ml/min) with: 1)
water; 2) ethylene oxide; 3)
propylene oxide; 4) ethylene
glycol; 5) propylene glycol.

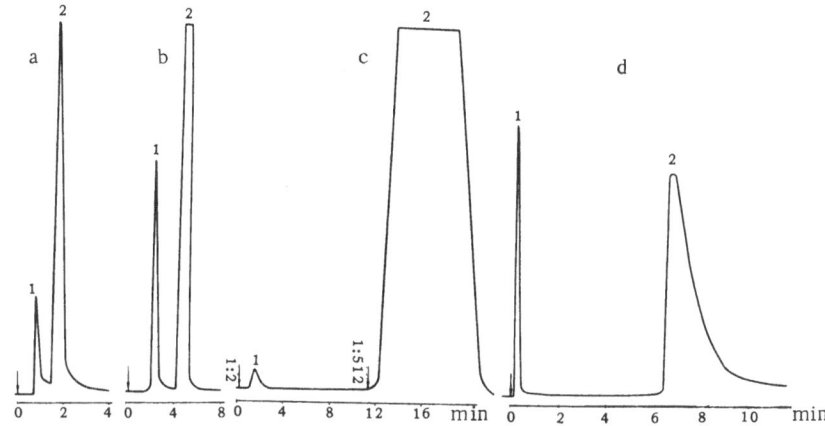

Fig. 168. Chromatograms on a Chromosorb 102 column: a) 1) ammonia; 2) water, 60°C; b) 1) water; 2) ethanol, 120°C; c) 1) water; 2) benzene, 120°C; d) 1) water; 2) glycerol, 150°C. Helium 30 ml/min; columns a-c) 300 × 0.3 cm, column; d) 200 × 0.3 cm.

Kelker [201] also examined liquid crystals of p, p'-azoxyphenetol, which gives complete separation of the isomeric xylenes in the anisotropic region (Fig. 161) (range 138 to 168°C).

Eutectics of inorganic salts on solid supports have also been used as sorbents in gas chromatography. Juvet and Wachi [203, 204] found no suitable liquid among those usually employed for separating volatile halides, but they found that their mixture of $SbCl_3$ with $TiCl_4$ could be separated on a eutectic melt of $BiCl_3$ with $PbCl_2$ (m.p. 217°C) at 240°C (Fig. 162). They also examined eutectics of KCl with $CdCl_2$ (m.p. 380°C) and $AlCl_3$ with NaCl (m.p. 183°C), but the last could not be used, because $AlCl_3$ is too volatile; also, its very strong acceptor interaction is a disadvantage in general use.

Hanneman et al. [205] found that the t_R for heterocyclic compounds on eutectics are more dependent on the structure than on the boiling point. Hanneman [206] made a detailed study of this with a eutectic mixture of the nitrates of Na, K, and Li (weight ratio 12.2 : 54.5 : 27.3) mounted on firebrick; he measured the V_R for picolines and lutidines at 150°C, which appear [206] to form complexes with the salts. Here there was no definite relation of V_R to boiling point; the molecular geometry largely controls the interaction. If the hydrogen in the α-position with respect to nitrogen is replaced by a methyl group, t_R is reduced; all the methylpyridines with α-substitution had t_R less than for β and γ-picolines. Substitution in the α-position on the other side of the nitrogen reduced t_R even more, which is obviously due to reduced accessibility of the N atom. This adsorbent was capable of separating the isomeric nitrogen bases.

Scott and Phillips [207] reported some important aspects of gas-adsorption chromatography, and, in particular, that this method gives chromatograms fully comparable with those from gas—liquid chromatography, with equal or better selectivity.

A very powerful method of analysis of mixtures is to combine displacement gas-adsorption chromatography with elution gas-adsorption or gas— liquid chromatography.

Displacement gas-adsorption chromatography is very promising for preparative work. Scott and Phillips have shown that symmetrical peaks (linear adsorption isotherms) occur not only for weakly adsorbed substances when modified solids are used. For instance, benzene on alumina gel modified with NaI ($Q \approx 15.7$ kcal per mole) is eluted with a symmetrical peak below 75°C. The properties of the surface are adjustable by choice of the salt, which provides greater scope for selectivity adjustment than in the case of gas—liquid chromatography; this, with the high thermal stability, makes such columns particularly suitable for many lengthy analyses.

Figure 163 shows hydrocarbons on a 90 cm column with particles of 100/150 mesh, the height equivalent to a theoretical plate being 0.35 mm. The thermal stability makes such columns particularly suitable for use with program temperature control; Fig. 164 shows an analysis of kerosene on a 150×0.3 cm column with temperature variation from 50 to 150°C.

Figure 165 shows that t_R for benzene relative to normal hydrocarbons increases when NaI replaces NaCl (see also Table 5 in Chapter II). The selectivity of these columns for aromatic compounds is higher than that of many liquid phases.

Inorganic salts have been used [262] either without a support or on support adsorbents with $s > 100$ m^2/g, as well as on supports of small s. The last gave the best results (a support for gas—liquid chromatography). The effects of the carrier on the separation have been discussed.

Chemisorption of some of the components of the mixture can be used to advantage on occasion. Adsorbents modified with copper chloride or silver nitrate adsorb alkenes irreversibly. Figure 166 shows curves given by $C_{10}-C_{14}$ alkanes and alkenes on alumina-gel columns modified with NaOH and NaOH + $CuCl_2$ [207]. In the second case the alkenes do not emerge from the column; the method can thus be used to remove alkenes from the mixture.

The selectivity of a solid adsorbent for cis and trans-isomers is usually higher than that of a liquid phase. The preparation of such columns is reproducible to 5% or better.

Components strongly adsorbed by active solids may be recovered by displacement chromatography. Combination of displacement with development in gas adsorption allows one to detect 0.01 ppm, e.g., for traces of hept-1-ene in heptane [207].

Displacement gas-adsorption chromatography is very promising for preparative purposes [207]; up to 50 g may be used per batch with a preparative gas-adsorption column.

Scott has used solid organic substances as adsorbents, in particular benzophenone on active alumina [208].

Ion-exchange resins modified with 1.5% squalane have been proposed [209] as adsorbents; C_6-C_9 fatty acids have been separated in this way, the peaks being almost symmetrical. This is probably a case of gas chromatography on monolayers (p. 53, Chapter II).

Some very promising weakly specific adsorbents are provided by porous polyaromatic polymers [210] derived from styrene or t-butylstyrene with divinylbenzene as cross-linking agent. Many different mixtures, such as oxides of nitrogen and glycols (Fig. 167), have been separated in this way. Nonspecific and weakly specific porous polymers are especially promising for detecting traces of water in solvents, hygroscopic liquids, and azeotropic mixtures [263] (Fig. 168). Porous polymers are also of interest in analysis of effluents for certain compounds. Water passes rapidly through the column and does not interfere with the separation of the organic compounds.

ANALYSIS OF HIGH-BOILING LIQUIDS

AND OF SOLIDS

Gas chromatography is now widely used at high temperatures [211], and extension of the temperature range gives good prospects for the analysis of high-boiling petroleum products (tars [212], lubricating oils [213]), fatty acids [188, 214], fatty-acid esters [215], plasticizers [216, 217], waxes [218], steroids [219], and even alloys [220]. Other possible applications are to the pyrolysis products from polymers, research on the thermal stability of adsorbents and catalysts, determination of organic substances in soils, research on catalysis and kinetics, study of inorganic reaction mechanisms, determination of oil in shales, preparation of small quantities of semiconductors of very high purity (preparative columns), and other applications to heavy products from synthesis or pyrolysis.

There are various difficulties in extending the temperature range upwards, apart from difficulties of apparatus and materials; these arise from the instability of the stationary phases.

A major problem for high-temperature chromatography is the preparation of adsorbents and supports. Most liquid phases are volatile or unstable at high temperatures, or even both simultaneously [211]. Moreover, the gas–liquid partition coefficient has a large negative temperature coefficient. Some possible high-temperatures sorbents are as follows [221]: polyamide resins, salt eutectics, and metal stearates, but most of these cannot be used above 500°C [221].

The most promising stationary phases for high-temperature chromatography are nonporous or macroporous inorganic materials, especially graphitized carbon black and macroporous silica gel.

Macroporous silica gel can be kept at 900°C for long periods without change in characteristics if it has previously been treated at higher temperatures.

Results have been reported [222] on the use of macroporous silica gel for separating high-boiling liquids and paraffin waxes; good separation was produced (Fig. 169a) for the normal hydrocarbons in paraffin wax, and also (Fig. 169b) for high-boiling phthalates often used as stationary phases in gas–liquid chromatography.

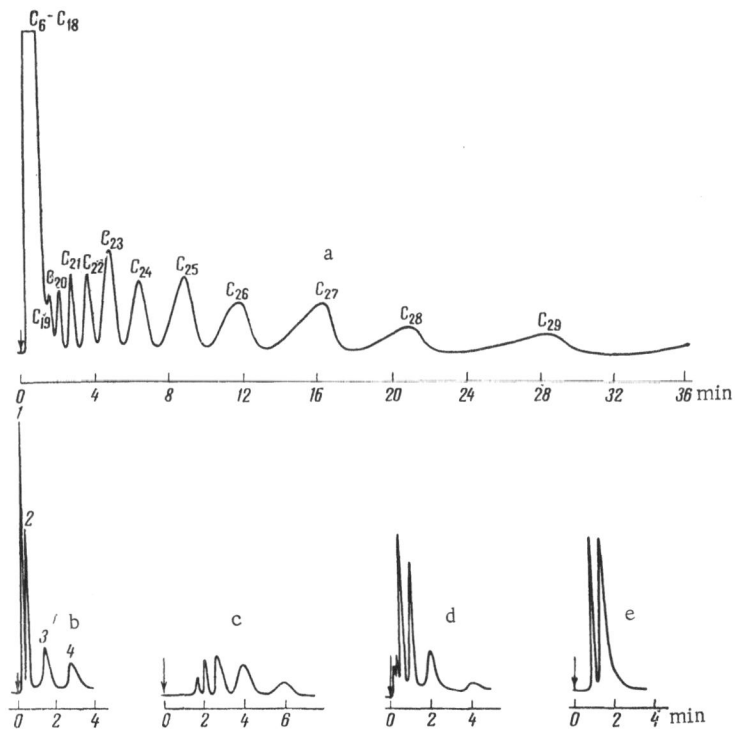

Fig. 169. Chromatograms on macroporous silica gel: a) $C_{19}-C_{29}$ normal hydrocarbons at 260°C; b) 306°C with: 1) dimethyl phthalate; 2) dibutyl phthalate; 3) dioctyl phthalate; 4) dinonyl phthalate; c) No. 5 silicone fluid at 332°C; d) VKZh-94 silicone oil at 332°C; e) 220°C with methyl esters of: 1) palmitic acid; 2) stearic acid.

The relation of log t_R (corrected) to 1/T gave the Q for these phthalates on this gel. Silicone oils No. 5 and VKZh-94 have been examined (Fig. 169c and d); these are also used for the same purpose, as well as for modifying silicates, glass materials, fillers, and as bonding agents in the preparation of paints. Figure 169c shows results for methyl palmitate and stearate. The above results are for gels with s of 7 and 20 m²/g (mean pore sizes respectively 4000 and 2500 Å).

These results show that macroporous silica gel is of general use in high-temperature gas chromatography of high-boiling organic substances, at least when these are of low polarity; in particular,

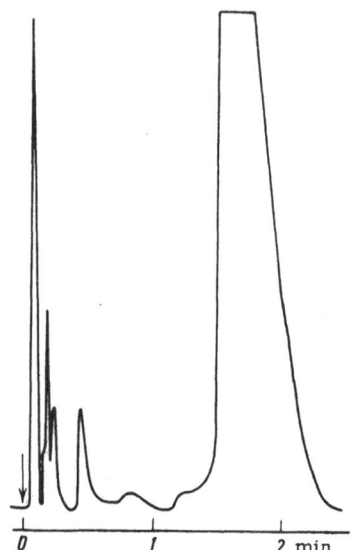

Fig. 170. Technical dinonyl phthalate at 280°C on 100 × 0.5 cm column of macroporous silica gel (s ≈ 20 m²/g).

they may be used in testing for impurities in the liquids used as stationary phases in gas–liquid chromatography, which is of particular importance if sensitive ionization detectors are to be used (Fig. 170). Such adsorbents allow the use of program temperature control, as in the separation of normal hydrocarbons.

Scott [208] examined the conditions giving symmetrical peaks with liquid hydrocarbons on alumina gel modified with NaOH when the working temperature was below the boiling point. This adsorbent can be used to separate hydrocarbons up to C_{45} (boiling point 350°C at 1 mm Hg) at 425°C [208].

Solomon [223] showed that inorganic salts may be used at high temperatures; Chromosorb treated with LiCl, CsCl, or $CaCl_2$ may be used at 200-500°C to separate the mixture of polyphenyls used as a coolant in nuclear reactors (Fig. 171); the separation of the isomeric tetraphenyls and hexaphenyls is nearly complete. The high boiling points of these substances (> 400°C) rule out the use of gas–liquid chromatography. Even the eutectics of [205] gave an unstable baseline when used with a temperature program from 200 to 500°C [223]. Quantitative analysis of the tetraphenyls and hexaphenyls was based on Chromosorb P (previously fired at 700°C) bearing 20% LiCl. This LiCl column can also be used with haloaromatics, aromatic ketones, and aromatic amines.

LiCl–Chromosorb P columns have also been used to separate the polynuclear aromatic carbons from indene to coronene (boiling points from 182 to 600°C) with the temperature increasing from 80 to 320°C at 2 deg/min (Fig. 172) [224]. The separation of isomeric benzopyrenes on such a column is as good as that on a capillary column equivalent to a much larger number of theoretical plates.

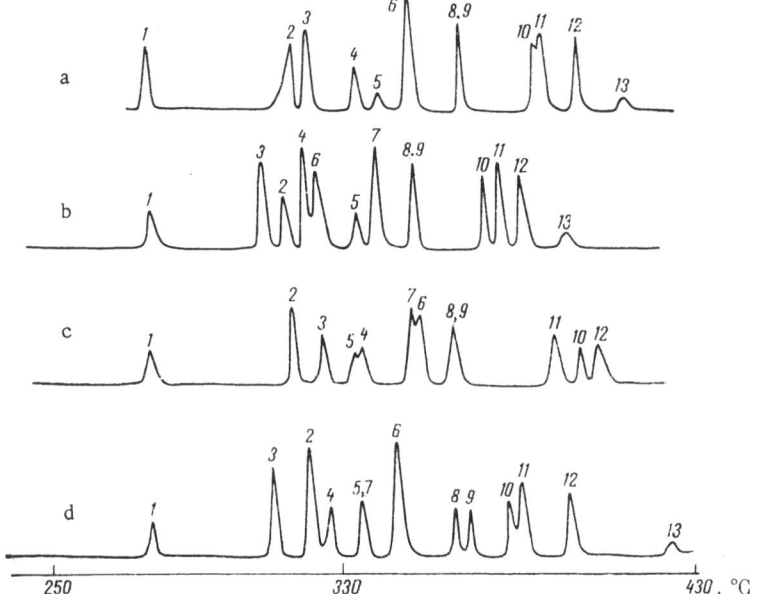

Fig. 171. Polyphenyls on a 300 × 0.6 cm Chromosorb column (He 200 ml/min) bearing: a) 20% LiCl; b) 20% CsCl; c) 50% CaCl₂; d) 20% CaCl₂ + CsCl; 1) o-terphenyl; 2) o-tetraphenyl; 3) m-terphenyl; 4) p-terphenyl; 5) 1,2,3-triphenyl-benzyl; 6) triphenylene; 7) o,m-tetraphenyl; 8) 1,2,4-triphenylbenzene; 9) o,p-tetraphenyl; 10) 1,3,5-triphenylbenzene; 11) m-tetraphenyl; 12) m,p-tetra-phenyl; 13) p-tetraphenyl.

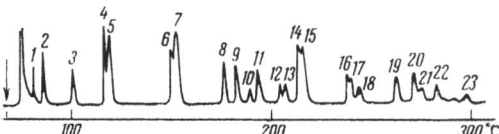

Fig. 172. Polynuclear aromatic hydrocarbons on a 240 × 0.3 cm column of Chromosorb P with 20% LiCl used with temperature variation (2 deg/min): 1) naphthalane; 2) indene; 3) 2-methylnaphthalene; 4) 2,6-dimethylnaphthalene; 5) acetonaphthalene; 6) 9-methylfluorene; 7) anthracene + phenanthrene; 8) fluoranthrene; 9) pyrene; 10) 9,10-dimethylanthracene; 11) 2-methylpyrene; 12) benzo(b)fluorene; 13) benzo(c)-fluroene; 14) benzo(b)phenanthrene + triphenylene; 15) benzo(a)anthracene + benzo(b)anthracene + chrysene; 16) benzo(e)pyrene; 17) benzo(a)pyrene; 18) benzo(b)fluoro-anthracene; 19) 3-methylcholanthene; 20) dibenzo(a,c)-anthracene and dibenzo(a)anthracene; 21) dibenzo(a, h)-anthracene; 22) coronene; 23) dibenzo(a, i)pyrene.

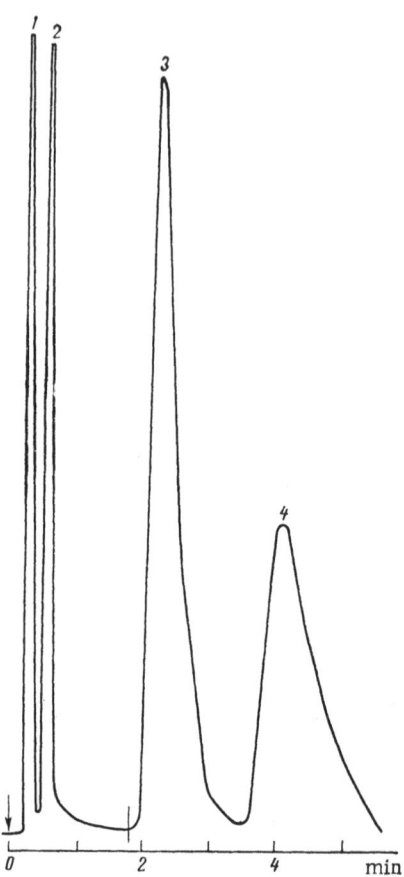

Fig. 173. Graphitized thermal carbon black
column 20× 0.2 cm at 360°C separating: 1)
benzene (solvent); 2) o-terphenyl; 3) m-terphenyl; 4) p-terphenyl (s = 7.6 m^2/g, N_2 30
ml/min, flame-ionization detector).

     Graphitized carbon black has been used to separate these poly-
nuclear aromatics [225]; the high selectivity of this material for
geometrical isomers allows the use of short columns. For example,
a 20 × 0.2 cm column at 360°C completely separated m-, o-, and
p-terphenyl (Fig. 173) [225, 261]. There is no relative twisting of
the phenyl rings in the p-position [226], so p-terphenyl emerges
last; the other two isomers have some rotation and so are well
separated from p-terphenyl.

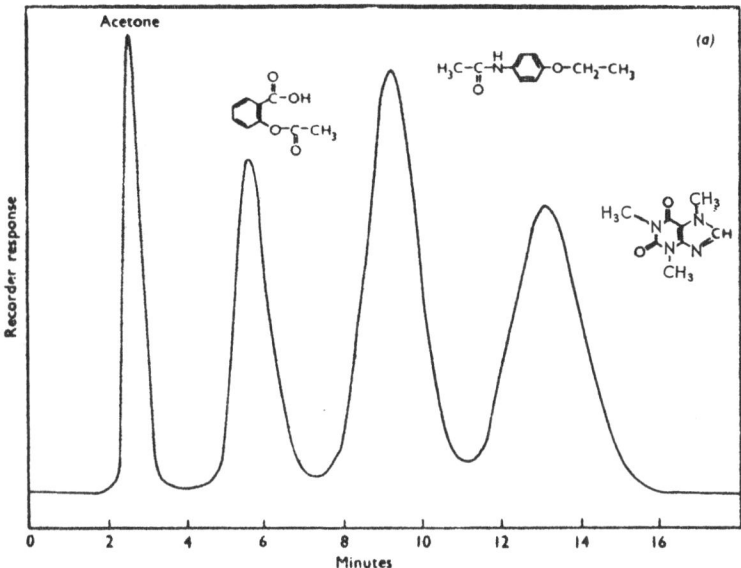

Fig. 174. Separation of acetylsalicyclic acid, phenacetin, and caffeine at 43 kg/cm^2 on a column 100 × 0.5 cm with the carrier diethyl ether at 200°C.

Macromolecular compounds of low volatility have been analyzed [264] by gas-adsorption chromatography by operation at pressures up to 50 atm with a carrier in a supercritical state. The adsorbent may be alumina gel [264] or Porapak Q [265, 266], while the carrier may be pentane (T_{cr} 196.62°C), diethyl ether (193.61°C), or isopropanol (235.25°C). This method has been used with a porous polymer to provide good separation of antioxidants, alizarin dyes, alkaloids, and epoxide resins [266]. Here the t_R vary with pressure and with the nature of the carrier. Phenol, hydroquinone, and phloroglucinol have been separated. The interaction with a polymer increases with the number of OH groups; phenols with three OH groups are eluted as symmetrical peaks. Pyrogallol has been separated from phloroglucinol (an isomer of the former), since here the separation is provided by the difference in molecular geometry and charge distribution. It has also proved possible to separate quinones and various inhibitors of oxidation or polymerization; these compounds emerge in order of increasing molecular weight. Figure 174 illustrates applications in pharmaceutical analysis, namely separation of the components of aspirin tablets (acetylsalicylic acid, phenacetin, and caffeine) and separation of cinchonine and brucine. The method is applicable to compounds of molecular weight above 1000 (Fig. 175).

An attempt has been made [220] to separate Zn and Cd by gas chromatography with LiCl on sea sand at 620°C as stationary phase, this being a little above the melting point of LiCl (614°C). The emerging metal vapor was condensed in a glass trap and analyzed spectroscopically.

Capillary gas-adsorption columns coated with alumina gel have been used in high-temperature separation of normal hydrocarbons [227]; in particular, the $C_{20} - C_{36}$ hydrocarbons were separated by programming the temperatures up to 400°C (Fig. 176).

ASPECTS OF IDENTIFICATION
AND QUANTITATIVE ESTIMATION

General Methods of Identification
in Gas Chromatography

Gas chromatography is mostly used as a quantitative method, the qualitative composition of the sample being already known; but qualitative analysis is required in certain cases.

The commonest detectors (katharometer, flame ionization, β-ionization) cannot be used for identification; but detectors have been described that allow direct identification at the exit. The most important of these is the time-of-flight mass spectrometer [228]. Some ionization detectors (in particular, the cross-section detector) can also give information on the nature of the compound [229]. Density indicators can give information on the molecular weight [230], while IR and UV spectrometers [231] or measurement of the velocity of sound [232] can give direct identification.

The time-of-flight mass spectrometer is usually employed in parallel with a katharometer, which performs the quantitative measurement. Here the mass spectrum is presented on an oscilloscope (which may be photographed), which can accept mass-to-charge ratios within the limits 1 to 6000.

Lovelock and Lipsky [233] described an ionization detector

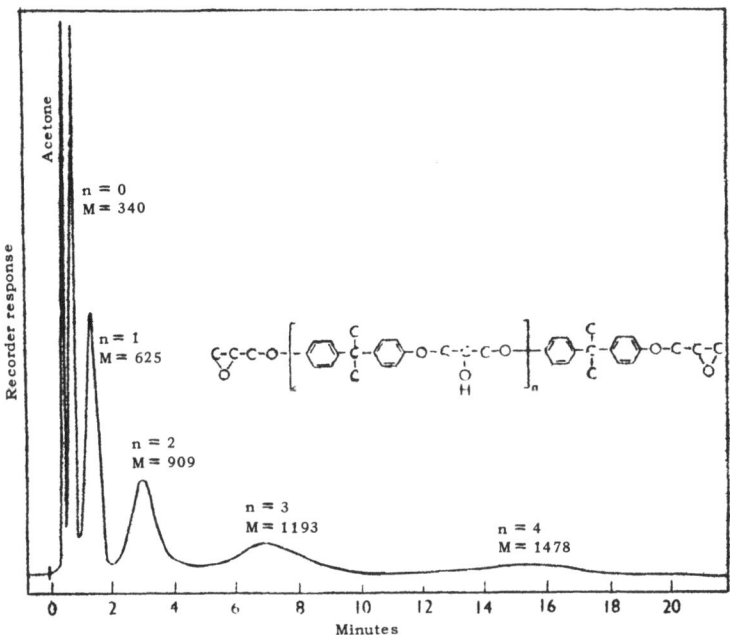

Fig. 175. Separation of the components of an epoxide resin of bisphenol-A-epichlorhydrin type on a column 25 cm × 6 mm at 200°C, carrier diethyl ether at 50 kg/cm².

that measures the electron affinity, which allows the detection of certain functional groups. The energy released by capture of a free electron serves to define the affinity. Many compounds (e.g., alcohols, ethers, esters, and ketones) differ widely in electron affinity.

Identification via a Martin—James density balance [234] was first described by Liberti et al. [230]; an internal standard of known molecular weight is added to the unknown component, and two chromatograms are run under the same conditions with different carrier gases, whereupon the molecular weight of the unknown component can be calculated.

Identification is simplified if the unknown component can be trapped at the exit; then instrumental methods can be combined with chemical analysis. Walsh and Merritt [235] describe various specific reactions for this purpose.

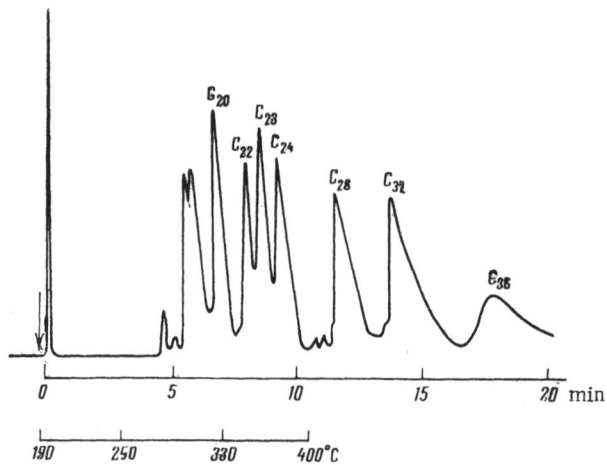

Fig. 176. $C_{20}-C_{36}$ hydrocarbons on a capillary column coated with alumina gel and used with program temperature control.

Chemical reactions are also used to remove certain classes of compound and thus to identify them [236], especially unsaturated compounds [237].

Molecular sieves (CaA zeolites) are widely used to remove normal saturated hydrocarbons, which are readily adsorbed, whereas geometrical factors prevent the adsorption of isoparaffins, cycloparaffins, and aromatic hydrocarbons.

Elementary analysis [238] (combustion to CO_2 and H_2O, with determination of the amounts of these) can also provide information on compounds.

Davison et al. [239] and Janak [240, 241] have used pyrolysis chromatography to establish the compositions of some nonvolatile organic compounds. At some given temperature, without access to air, the compound (especially a polymer) splits up into a series of compounds indicating its composition.

Chromatography itself often provides identification, whatever the detector, since t_R or V_R is dependent on the nature of the compound. Addition of a known compound or comparison of V_R provides identification, though the result is not always unambiguous.

Identification within a homologous series is facilitated by the linear relation of $\log t_R$ or $\log V_R$ to the number of carbon atoms in the molecule [162, 242]; Q is also dependent on that number for straight-chain compounds [243] (Chapter II, Fig. 7).

The relation of log t_R to a dimensionless parameter (ratio of boiling point to working temperature) has been used [37] in the identification of nonpolar substances. Wehrli and Kovats [244] have used the retention index I, which is defined by:

$$I = 200 \frac{\log[(t_R)_i/(t_R)_{p_z}]}{\log[(t_R)_{p_{z+2}}/(t_R)_{p_z}]} + 100z, \qquad (V, 1)$$

in which $(t_R)_i$ is for substance i, $(t_R)_{p_z}$ is the retention time for normal alkane p_z (containing z carbon atoms), and $(t_R)_{z+2}$ is the same for p_{z+2} (two carbons atoms extra), these two alkanes being chosen to be such that $(t_R)_{p_z} \leq (t_R)_i \leq (t_R)_{p_{z+2}}$.

Some Special Methods of Identification
in Gas-Adsorption Chromatography

We have seen (p. 20) that the V_S are reproducible physico-chemical constants dependent on the nature of the system at given T and θ; hence the V_S, and also these relative to the corresponding quantities for the normal alkanes, may be used to identify components from macroporous and nonporous adsorbents, e.g., via the use of the I above.

The relation of log V_S to n (number of carbon atoms) has [243] been recorded for several homologous series; the relation is always linear. Also, Q is linearly related to n for the normal alkanes and their derivatives (Fig. 6), which can be used in identification. The Q for normal alkanes have been determined for silica gel up to C_{25} and for graphitized carbon black up to C_{17} (Fig. 35).

Nonspecific adsorbents give high selectivity for structural and geometrical isomers [180, 181], and the V_R give some information on the geometrical structure. This provides a means of conformational analysis for unknown isomers if the conformations are stable.

Quantitative Analysis

This is based on the proportional relation of peak area to amount of concentration of the component; peak height, the product of height by V_R, and the product of height by t_R are also used. Peak areas are measured in various ways, in accordance with their size, symmetry, and completeness of separation. The commonest methods are [245] as follows:

1. From the area of the triangle formed by the baseline and the tangents at the points of inflection.

2. Multiplication of the peak height h by the width $\Delta t_{1/2}$ measured at half height:

$$A = h\Delta t_{1/2},$$

(V, 2)

(here it is assumed that the triangle is approximately isosceles);

3. Weighing after cutting-out of the peak.

4. By planimeter.

5. By electromechanical integrator during recording.

6. By electronic integrator during recording.

The following formulas are used for the areas when the separation is incomplete:

$$A = 2.507h\sigma,$$

(V, 3)

in which h is height and σ is the standard deviation (width Δt measured at height 0.882h; $\Delta t = 2\sigma$ at height 0.607h and $\Delta t = 3\sigma$ at height 0.324h) and

$$A = kh\Delta t,$$

(V, 3a)

in which Δt is width at one of the distances 0.5h, 0.75h, or 0.9h and k is a correction factor (1.065, 166, or 2.73, respectively).

A statistical comparison [245] of these various methods shows that the electronic integrator gives the least error (down to 1.7%), though the errors of all methods are large (15-25%) if the areas are small (under 50 mm^2). In the latter case, with narrow peaks, the best results are given by the use of peak heights.

The response of a detector varies with a homologous series, as well as between such series, so the peak areas will not be equal

even if the mixture contains all components in the same proportion. Correction factors (relative sensitivities) must be used to
reduce peak areas to quantities proportional to the concentrations.

Three methods are most commonly used [246-248]; absolute
calibration, internal standards, and use of correction factors in
normalization.

Absolute Calibration: The amount of the substance
is deduced from a calibration curve for the area or peak height as
a function of amount used; pure substances must be used in the calibration, and the operating conditions of the columns must be identical, especially when peak heights are used. This method is most
commonly employed when there is no need to determine all the
components, and it may be recommended when the gas samples are
reasonably large, since the error in determining the sample size
is then small; the error is greater for liquid samples, so it is best
to use dilute solutions of accurately known concentration.

Internal Standard: A known amount of some substance
is added (internal standard), the correction factor being taken as
one for this. A graph is drawn up for the ratio of the areas (or
heights) given by a compound and the internal standard as a function of the ratio of weights or percent contents. This graph is drawn
up from a series of chromatograms with different values of the latter ratio. Then a known amount of the internal standard is added
to the sample; the ratio of areas or heights is determined, and hence
the concentration of the compounds is read from the graph. This is
the most convenient and widely used method.

In this case there is no need to determine the size of the sample exactly, which is one of the main advantages, since exact volume measurement on small liquid or gas samples is difficult, especially with manually operated syringes. The method also eliminates effects from change of gas speed or column temperature.
However, it is used only when it is not necessary to assay all the
components in the sample.

Normalization: Here the sum of all peak areas, with
the correction factors incorporated, is taken as 100%; the percent
content of any given substance is then given by simple relationships.
One advantage is that the percent content can be determined for
all components.

COMPARISON OF GAS-ADSORPTION

AND GAS – LIQUID METHODS

FOR IDENTICAL MIXTURES

It is generally accepted [37] that gas adsorption is better for gases and low-boiling compounds, so no comparison is needed for these. There are, however, various other important cases in which it is of interest to make the comparison.

First we consider high-boiling mixtures. At present there are no liquid phases that can be used above 300°C when an ionization detector is employed; even apiezon oils and silicone elastomers above 250°C give backgrounds so high that sensitive detectors cannot be used in isothermal working. Figure 177 shows that the background ion currents given by apiezon L and E-301 silicone elastomer rise very rapidly above 250°C. This rise in background current is accompanied by an increase in the fluctuations, which are

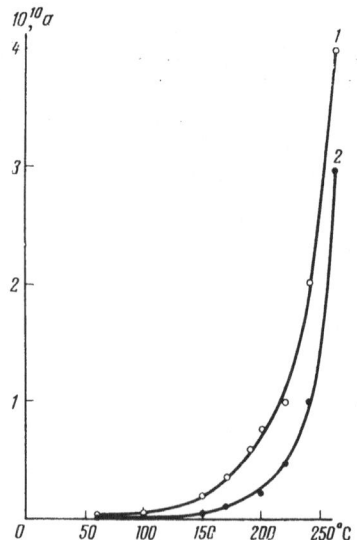

Fig. 177. Background current of a flame-
ionization detector as a function of temperature for gas-liquid columns containing: 1) E-301 silicone elastomer; 2)
apiezon L.

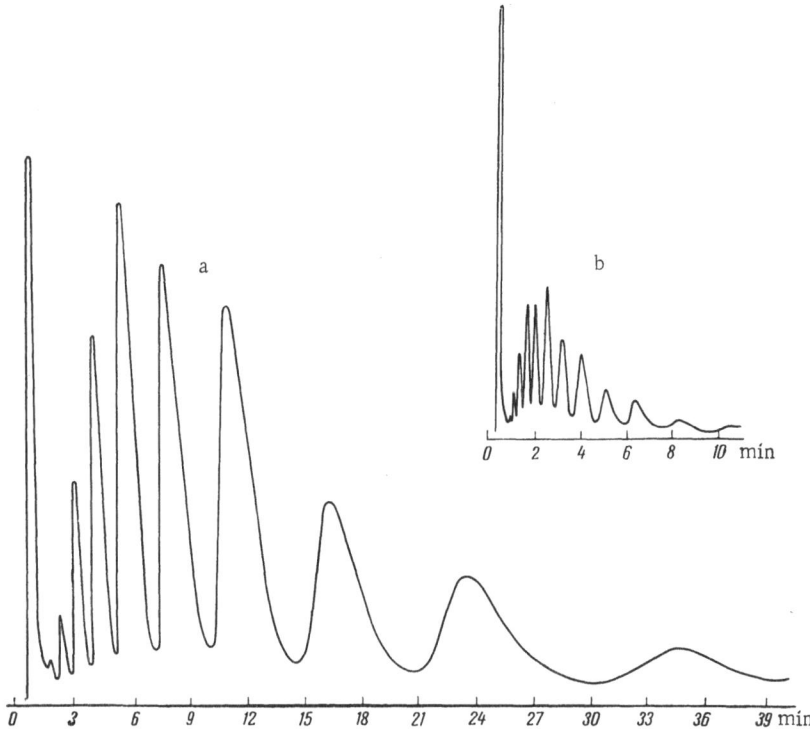

Fig. 178. $C_{19}-C_{29}$ normal alkanes on 100×4 cm columns with flame-ionization detector: a) macroporous silica gel ($s \approx 26$ m^2/g, 250°C); b) silicone elastomer (15% on diatomite brick, 280°C).

difficult to balance out. There are thus obvious advantages in gas-adsorption chromatography with ionization detectors above 300°C. Again, gas-adsorption chromatography is comparable with gas—liquid chromatography as regards selectivity and performance at 200-300°C, as Fig. 178 shows for the $C_{19}-C_{29}$ normal alkanes at 250°C on macroporous silica gel (prepared by the method of [249]) and on E-301 supported by diatomite brick. Figure 179 also shows that gas-adsorption chromatography more readily allows the use of temperature programming.

Gas adsorption is preferred for determination of trace components with maximally sensitive ionization detectors, e.g., at 0.01-0.1 ppm it is impossible to use gas—liquid columns [146].

Gas adsorption may be of higher performance at any temperature, since the mass exchange with the surface of a nonporous or

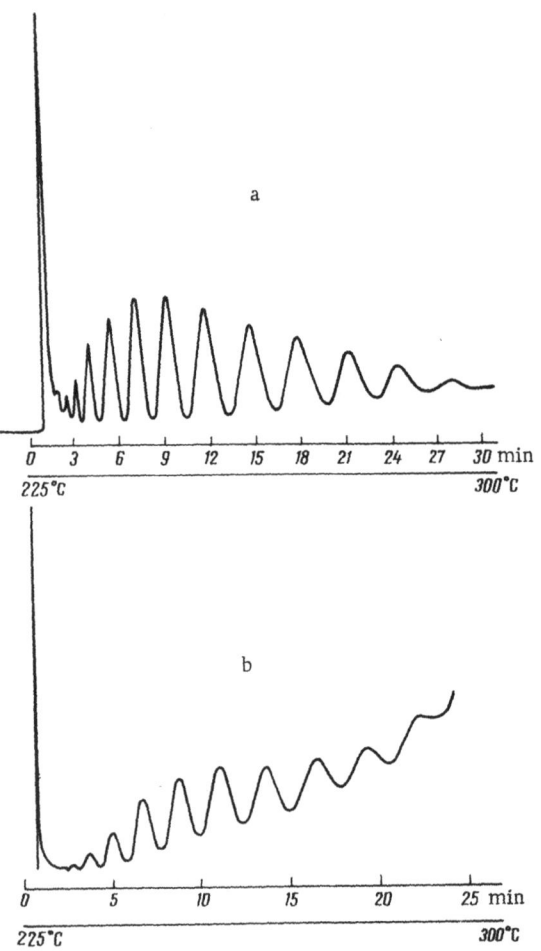

Fig. 179. C_{19}—C_{29} normal alkanes used with program temperature control on 100×4 cm columns containing: a) macroporous silica gel (s $\approx$ 26 m^2/g); b) apiezon L (15% on diatomite brick).

macroporous solid giving nonspecific or weakly specific adsorption is much more rapid than the transfer within the solution in the liquid [162]; hence gas adsorption allows many more uptake—release cycles to occur in unit time, which gives a column with greater performance and a larger number of theoretical plates [250]. The peak broadening in gas adsorption may be minimized by the use of adsorbents with maximally uniform surfaces in the form of equal spheres uniformly packed in the column [124]. This can give a

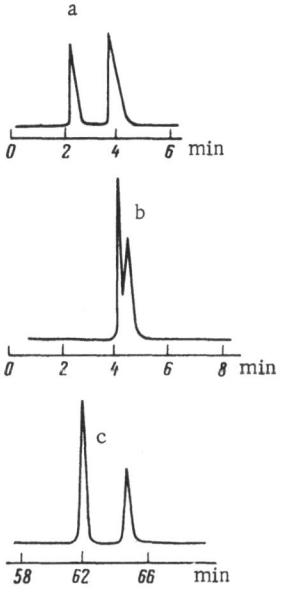

Fig. 180. Mixture of trans and cis isomers of normal 1-methyl-3-butylcyclohexane on columns of: a) graphitized carbon black 100 × 0.4 cm (s = 7.6 m²/g) at 216°C (N = 700); b) capillary coated with polypropylene glycol at 109°C (N = 50,000); c) capillary coated with squalane at 90°C (N = 110,000).

column in which the height equivalent to one theoretical plate is less than 0.5 mm, which provides high performance.

The selectivity in adsorption is also often higher than that in solution, so an adsorption column equivalent to a relatively small number of theoretical plates gives better separation than a packed gas–liquid column or even an open liquid-coated capillary. In particular, gas adsorption is much more selective for geometrical isomers. For instance, the cis and trans isomers of alkylcyclohexanes (Fig. 180) are well separated on a graphitized carbon black column equivalent to only 700 theoretical plates, whereas equal separation on a capillary gas–liquid column requires the equivalent of some tens of thousands of plates.

Other adsorbents are inorganic salts [196, 223] and modified oxides [207], as well as nonporous or macroporous adsorbents bearing close-packed monomolecular adsorbed layers of high-boiling liquids or unbranched polymers containing a variety of functional groups. These approaches give control of the surface chemical composition and of the balance between nonspecific and specific interactions (Chapter II); they thus provide selectivity control for gas-adsorption columns.

The rapid mass transfer in gas-adsorption chromatography can be used to advantage in high-speed analysis, especially by the use of nonporous or surface-porous adsorbents [162].

Although gas-adsorption chromatography has advantages over gas–liquid chromatography in these cases and many others, the two methods will undoubtedly develop in parallel, especially for many

applications in the range 50-200°C. It will be clear from the above
that gas-adsorption chromatography has been unduly neglected in
the past, as has the necessary development work on adsorbents.

LITERATURE CITED

1.   H. P. Burchfield and E. E. Storrs, Biochemical Applications of Gas Chro-
      matography, New York, Academic Press, 1962, p. 193.
2.   M. Kokuba, T. Mayeda, and H. C. Urey, Geochim. Cosmochim. Acta,
      21:247 (1961).
3.   A. J. P. Martin and R. L. M. Synge, Biochem. J., 35:1358 (1941).
4.   E. Gleuckauf and G. P. Kitt, in collection: Proceedings of the International
      Symposium on Isotope Separation, New York, Interscience Publishers Inc.,
      1958, p. 210.
5.   C. O. Thomas and H. A. Smith, J. Phys. Chem., 63:427 (1959).
6.   S. Ohkoshi, Y. Fujita, and T. Kwan, Bull. Chem. Soc. Japan, 31:770 (1958).
7.   O. Riedel and E. Uhlman, Z. Anal. Chem. 166:433 (1959).
8.   E. M. Arnett, M. Strem, N. Hepfinger, J. Lipowitz, and D. M. Guire, Science,
      131:1680 (1960).
9.   W. R. Moore and H. R. Ward, J. Am. Chem. Soc., 80:2909 (1958); J. Phys.
      Chem., 64:832 (1960).
10.   H. A. Smith and P. P. Hunt, J. Phys. Chem., 64:383 (1960).
11.   M. Venugopalan and K. O. Kutschke, Can. J. Chem., 41:548 (1963).
12.   W. A. van Hook and P. H. Emmett,   J. Phys. Chem., 64:673 (1960).
13.   G. Shipman, Anal. Chem., 34:877 (1962).
14.   S. Furuyama and T. Kwan, J. Phys. Chem., 65:190 (1961).
15.   T. Kwan, J. Res. Inst. Catalysis, Hokkaido Univ., 8:18 (1960); Chem. Ind.,
      14:723 (1961).
16.   P. P. Hunt and H. A. Smith, J. Phys. Chem., 65:87 (1961).
17.   F. Botter, G. Perriere, and S. Tistchenko, Comm. Energie At. (France),
      Rappt. C. E. A., No. 1962, 1961.
18.   L. Bachmann, E. Bechtold, and E. Cremer, J. Catalysis, 1:113 (1963).
19.   E. Cremer, Z. Anal. Chem., 170:219 (1959).
20.   E. Glueckauf, Z. Chim. Phys., 60:73 (1963).
21.   P. L. Cant and K. Jang, Science, 129:1548 (1959).
22.   Y. Fujita and T. Kwan, Japan Analyst, 60:15 (1963).
23.   J. L. Borowitz, Israel J. Chem., 1:378 (1963).
24.   D. L. West and A. L. Marston, J. Am. Chem. Soc., 86:4731 (1964).
25.   H. A. Smith and E. H. Carter, in collection: Proceedings of the International
      Atomic Energy Agency Symposium on the Use of Tritium in the Physical and
      Biological Sciences, Vol. 1, Vienna, 1962, p. 121.
26.   E. H. Carter and H. A. Smith, J. Phys. Chem., 67:535 (1963).
27.   E. H. Carter and H. A. Smith, J. Phys. Chem., 67:1512 (1963).
28.   M. Mohnke and W. Saffert, in collection: Gas Chromatography 1962, M. M.
      Swaay (ed.), London, 1962, p. 214.
29.   D. White and E. N. Lassettre, J. Chem. Phys., 32:72 (1960).

30. A. A. Evett, J. Chem. Phys., 31:565 (1959).

31. T. Phillips, D. Owens, and A. Hamlin, Nature, 192:1067 (1961).

32. K. I. Sakodynskii, in collection: Gas Chromatography, Moscow, Izd. NIITÉKhIM, Issue 2, 1964, p. 5.

33. S. Akhtar and H. A. Smith, Chem. Rev., 64:261 (1964).

34. C. Cercy and F. Botter, Bull. Soc. Chim. France, No. 11, 1965, p. 3383.

35. D. Ambrose and B. A. Ambrose, in collection: Gas Chromatography, London, George Newnes Ltd., 1961, p. 127.

36. J. H. Knox, in: Gas Chromatography, London, Methuen, 1962, p. 41.

37. A. A. Zhukhoviskii and N. M. Turkel'taub, Gas Chromatography, Moscow, Gostoptekhizdat, 1962.

38. V. A. Sokolov, Methods of Gas Analysis, Moscow, Gostoptekhizdat, 1958.

39. S. P. Zhdanov, V. I. Kalmanovskii, A. V. Kiselev, N. Ya. Smirnov, M. M. Fiks, and Ya. I. Yashin, Summaries of Papers at the Intercollegiate Conference on Adsorption and Chromatographic Analysis Methods, Odessa, 1961, p. 26.

40. S. A. Greene, M. L. Moberg, and E. J. Wilson, Anal. Chem., 28:1369 (1956).

41. G. Kuryacos and C. E. Boord, Anal. Chem., 29:787 (1957).

42. F. Rarre-Rius and G. Guiochon, J. Chromatog., 13:383 (1964); J. Gas Chromatog., 1:33 (1963).

43. E. W. Lard and R. C. Horn, Anal. Chem., 32:878 (1960).

44. A. J. Martin, C. E. Bennett, and F. W. Martinez, in collection: Gas Chromatography, H. J. Noebels, R. F. Wall, and N. Brenner, (eds.), New York, Academic Press, 1961, p. 363.

45. T. L. Thomas and R. L. Moys, Phys. Met. Chem. Anal., 4:45 (1961).

46. J. Takashima, A. Koga, and J. Kaneko, J. Chem. Soc. Japan, 65:1223 (1962).

47. S. A. Greene and H. Pust, Anal. Chem., 29:1055 (1957); 30:1039 (1958).

48. E. L. Szonntagh and J. R. Stewart, Symposium on Gas Chromatography, Third Delaware Regional Meeting of American Chemical Society, 1960.

49. G. S. Vizard and A. Wynne, Chem. Ind., London, 1959, p. 196.

50. G. W. Heylmun, J. Gas Chromatog., 3:82 (1965).

51. M. Kreiči, K. Tesařik, and J. Janak, in collection: Gas Chromatography, H. J. Noebels, F. R. Wall, and N. Brenner (eds.), New York, Academic Press, 1961, p. 255.

52. K. Abel, Anal. Chem., 36:953 (1964).

53. J. W. Swinnerton, V. J. Linnenbom, and C. H. Cheek, Anal. Chem., 36:1669 (1964).

54. E. L. Obermiller and R. W. Freedman, J. Gas Chromatog., 3:242 (1965).

55. A. V. Kiselev, Yu. L. Chernen'kova, and Ya. I. Yashin, Neftekhimiya, 5:141 (1965).

56. A. V. Kiselev, Yu. L. Chernen'kova, and Ya. I. Yashin, Neftekhimiya, 5:589 (1965).

57. N. M. Turkel'taub, O. V. Zolotareva, A. G. Latukhova, N. I. Karymova, and E. R. Kal'nina, Zh. Anal. Khim., 11:159 (1956).

58. W. M. Graven, Anal. Chem., 31:1197 (1959).

59. S. P. Zhdanov, A. V. Kiselev, and Ya. I. Yashin, in collection: Gas Chromatography, Moscow, Izd. NIITÉKhIM, Issue 1, 1964, p. 5.

60. G. Pannetier and G. Djega-Mariadassou, Bull. Soc. Chim. France, No. 7: 2089 (1965).
61. J. Janak, Collection Czech. Chem. Commun., 19:917 (1954).
62. J. Janak, Chem. Listy, 47:837, 1348 (1953).
63. S. A. Greene, Anal. Chem., 31:480 (1959).
64. M. Krejči, Collection Czech. Chem. Commun., 25:2457 (1960).
65. J. Janak, Ann. New York Acad. Sci., 72:666 (1959).
66. M. Wilkins and D. Wilkins, Rep. At. Energy Research Estab. U.S.A., 1960, p. 15.
67. J. Tothand and L. Graf, Magy. Kem. Folyoirat, 63:216 (1957).
68. N. M. Turkel'taub, L. N. Ryabchuk, S. N. Morozova, and A. A. Zhukhovitskii, Zh. Anal. Khim., 19:133 (1964).
69. E. Zielinski, Chem. Anal., 5:605 (1960).
70. Abolghassen Ghalamsiah, Comm. Energie At. (France), Rappt. C. E. A., No. 2361:86 (1963).
71. Ya. I. Yashin, Dissertation, Moscow University, 1965.
72. E. Glueckauf and G. P. Kitt, Proc. Roy. Soc., A294:557 (1956).
73. E. V. Vagin and S. S. Petukhov, Kislorod, No. 2, 1959, p. 33.
74. S. S. Petukhov and E. V. Vagin, in collection: Gas Chromatography, Moscow, Izd. Akad. Nauk SSSR, 1960, p. 292.
75. E. V. Vagin and S. S. Petukhov, Kislorod, No. 2, 1958, p. 44.
76. S. S. Petukhov and E. V. Vagin, in collection: Gas Chromatography, Moscow, GOSINTI, Issue 3, 1960, p. 27.
77. A. A. Zhukhovitskii and N. M. Turkel'taub, Usp. Khim., 25:859 (1956).
78. C. T. Hodges and R. F. Matson, Anal. Chem., 37:1065 (1965).
79. D. Y. Yee, J. Gas Chromatog., 3:314 (1965).
80. E. E. Neely, Anal. Chem., 32:1382 (1960).
81. J. A. Murdoch, Analyst, 86:856 (1961).
82. G. Nodop, Z. Anal. Chem., 164:120 (1958).
83. D. G. Timms, H. J. Konrath, and R. C. Chirnside, Analyst, Vol. 83 (1958).
84. A. W. Mosen and G. Buzzelli, Anal. Chem., 32:141 (1960).
85. R. Whatmough, Nature, 179:911 (1957).
86. K. Drekopf and W. Winzen, Glückauf, 93:1222 (1957).
87. L. Bovin, J. Pirott, and A. Berget, in collection: Gas Chromatography, Proceedings of the Second Symposium Organized by the Gas Chromatography Discussion Group at the Royal Tropical Institute, Amsterdam, and at the Conference on the Analysis of Mixtures of Volatile Substances in New York [Russian translation], Moscow, IL, 1961, p. 288 [English edition: D. H. Desty (ed.), New York, Academic Press, 1959].
88. A. A. Avdeeva, Teploénerg, No. 8, 1959, p. 16.
89. F. G. Druzhinina and V. P. Mityanin, Zavodsk. Lab., 30:531 (1964).
90. J. Lacy, K. G. Woolmington, and R. V. Hill, S. African Ind. Chemist, 14: 47 (1960).
91. A. Krhut, O. Kalanora, and L. Brhaček, Hutnicke Listy, 15:133 (1960).
92. E. Zielinski, Przeglad Odlewnictwa, 13:219 (1963); J. Kashima, T. Muraki, and T. Jamazaki, J. Japan Found. Soc., 32:40 (1960).

93.   F. M. Evens and V. A. Fassel, Anal. Chem., 35:1444 (1963); H. Feichtinger,
      H. Bächtold, and W. Schuhknecht, Schweiz. Arch. Angew. Wiss. Tech.,
      25:426 (1959).
94.   R. Y. Hynek and J. A. Nelen, Anal. Chem., 35:1655 (1963).
95.   P. G. Elsey, Anal. Chem., 31:869 (1959).
96.   J. Halasz and W. Schneider, Brennstoff-Chem., 41:225 (1960).
97.   A. A. Kilner and G. A. Ratelift, Anal. Chem., 36:1615 (1964).
98.   H. L. Ashmead, G. E. Martin, and J. A. Schmit, J. Assn. Off. Agr. Chem.,
      47:730 (1964).
99.   W. Kleber, A. Hartl, and P. Schmid, Brauwissenschaft, 16:106 (1963).
100. L. M. Hamilton and R. C. Kory, J. Appl. Phys., 15:829 (1960).
101. A. M. Dominguez, H. E.Christenson, L. R. Goldbaum, and A. Stembridge,
      Toxicol. Appl. Pharmacol., 1:135 (1959).
102. L. H. Ramsey, Science, 129:900 (1959).
103. D. H. Smith and F. E. Clark, Soil Sci. Soc. Am. Proc., 24:111 (1960).
104. J. Janak and M. Rusek, Chem. Listy, 48:397 (1954); D. H. Smith and F. E.
      Clark, Soil Sci. Soc. Am. Proc., 24:111 (1960).
105. R. P. De Grazio, J. Gas Chromatog., 3:204 (1965).
106. S. A. Greene and H. Pust, Anal. Chem., 30:1039 (1958).
107. L. Marville and J. Tranchant, in collection: Gas Chromatography(Trans-
      actions of the Third International Symposium on Gas Chromatography in
      Edinburgh) [Russian translation], Moscow, Izd. Mir, 1964, p. 413.
108. T. N. Gvozdovich and Ya. I. Yashin, in collection: Gas Chromatography,
      Moscow, Izd. NIITÉKhIM, Issue 6, 1967.
109. O. L. Hollis, Anal. Chem., 38:309 (1966).
110. F. Bruner and G. P. Cartoni, J. Chromatog., 18:390 (1965); F. Bruner, G. P.
      Cartoni, and A. Liberti, Anal. Chem., 38:298 (1966).
111. M. E. Morrison, R. G. Rinker, and W. H. Corcoran, Anal. Chem., 36:2256
      (1964).
112. M. Tsuchiga, V. Yasuda, and H. Komada, Japan Analyst, 14:155 (1965).
113. J. A. Barnard and H. W. E. Hughes, Nature, 183:24 (1959).
114. E. M. Fredericks and F. R. Brooks, Anal. Chem., 28:297 (1956).
115. O. Horn, U. Schwenk, and H. Hochenberg, Brennstoff-Chem., 88:116 (1957).
117. N. M. Turkel'taub, A. I. Kolbasina, and M. S. Semenkina, Zh. Fiz. Khim.,
      12:302 (1957).
118. H. W. Patton, J. S. Lewis, and W. I. Kaye, Anal. Chem., 27:170 (1955).
119. L. V. Polyakova, G. I. Sal'nikova, and Ya. I. Yashin, in collection: Gas
      Chromatography, Moscow, Izd. NIITÉKhIM, Issue 5, 1967.
120. T. A. McKenna and J. A. Idleman, Anal. Chem., 32:1299 (1960).
121. N. I. Lulova, A. I. Tarasov, A. V. Kuz'mina, and N. M. Koroleva, Nefte-
      khimiya, 2:885 (1962).
122. G. J. Taylor and A. S. Dunlop, Gas Chromatography, V. J. Coates, H. J.
      Noebels, and I. S. Fagerson (eds.), New York, Academic Press, 1958, p. 73.
123. D. A. Vyakhirev, M. V. Zueva, and A. I. Bruk, Papers on Chemistry and
      Chemical Technology, Gor'kii, Vol. 2, 1964, p. 268.
124. A. V. Kiselev, I. A. Migunova, and Ya. I. Yashin, Neftekhimiya, 7:807 (1967).

125.   V. Crespi and F. Cevolani, Chim. Ind. (Milan), 41:215 (1959).
126.   C. Rossi, S. Munari, and L. Congarle, Chim. Ind. (Milan), 42:724 (1960).
127.   S. P. Zhdanov, A. V. Kiselev, V. I. Kalmanovskii, M. M. Fiks, and Ya. I.
       Yashin, in collection: Gas Chromatography, Moscow, Izd. Nauka, 1964,
       p. 61.
128.   Ya. I. Yashin, S. P. Zhdanov, and A. V. Kiselev, Gas Chromatographie,
       1963, H. P. Angele and H. G. Struppe (eds.), Berlin, Akademie Verlag,
       1963, p. 402.
129.   S. P. Zhdanov, Dokl. Akad. Nauk, 82:281 (1952).
130.   A. A. Zhukhovitskii and N. M. Turkel'taub, Usp. Khim., 30:877 (1961).
131.   S. P. Zhdanov, A. V. Kiselev, and Ya. I. Yashin, Zh. Fiz. Khim., 37:1432
       (1963).
132.   M. S. Vigdergauz, K. A. Gol'bert, and O. R. Gorshunov, Khim. i Tekhnol.
       topliv i Masel, 7:62 (1961).
133.   I. Halasz and E. Heine, Nature, 194:971 (1962).
134.   A. G. Svyatoshenko and V. G. Berezkin, Neftekhimiya, 4:938 (1964).
135.   S. P. Zhdanov, A. V. Kiselev, V. I. Kalmanovskii, M. M. Fiks, and Ya. I.
       Yashin, Zh. Fiz. Khim., 36:1118 (1962).
136.   F. A. Bruner and G. P. Cartoni, Anal. Chem., 36:1522 (1964).
137.   R. D. Schwartz, D. J. Brasseaux, and G. R. Shoemake, Anal. Chem., 35:496
       (1963).
138.   D. L. Petitjean and C. J. Leftault, J. Gas Chromatog., 1:18 (1963).
139.   J. J. Kirkland, Anal. Chem., 35:1295 (1963).
140.   W. Schneider, H. Bruderreck, and I. Halasz, Anal. Chem., 36:1533 (1964).
141.   I. Halasz and H. O. Gerlach, Anal. Chem., 38:281 (1966).
142.   C. A. Gaulin, E. R. Michaelsen, A. B. Alexander, and R. W. Souer, Chem.
       Eng. Progr., 56:49 (1958).
143.   F. H. Huyten, G. W. A. Rijnders, and W. V. Beersum, in collection: Gas
       Chromatography 1962, M. van Swaay (ed.), London, 1962, p. 335.
144.   G. M. Sal'nikova and Ya. I. Yashin, in collection: Gas Chromatography,
       Moscow, Izd. NIITÉKhIM, Issue 4, 1966, p. 118.
145.   N. I. Lulova, A. I. Tarasov, A. V. Kuz'mina, A. K. Fedorova, and S. A.
       Leont'eva, in collection: Gas Chromatography, Moscow, Izd. Nauka, 1964,
       p. 162.
146.   C. J. Kuley, Anal. Chem., 35:1472 (1963).
147.   A. L. Roberts and C. P. Word, J. Inst. Gas Eng., 5:313 (1965).
148.   M. I. Dement'eva, T. I. Naumova, V. E. Shefter, and G. S. Fedchenko,
       in collection: Separation and Analysis of Hydrocarbon Gases, Moscow, Izd.
       Akad. Nauk SSSR, 1963, p. 142.
149.   H. Miyake, J. Japan Chem., 12:627 (1958); T. A. Mikenna and J. A. Idle-
       man, Anal. Chem., 32:1299 (1960).
150.   W. H. Heaton and J. T. Wontworth, Anal. Chem., 31:349 (1959).
151.   V. Veber and N. M. Turkel'taub, Geol. Nefti, No. 8, 1958, p. 39.
152.   V. A. Sokolov and N. M. Turkel'taub, in collection: Geochemical Meth-
       ods of Exploration for Oil and Gas, Moscow, Gostoptekhizdat, Issue 1, 1953,
       p. 3.

153.   A. V. Kiselev, Vestn. Mosk. Gos. Univ., Ser. Khim., No. 5, 1961, p. 31.
154.   N. V. Akshinskaya, V. E. Beznogova, A. V. Kiselev, and Yu. S. Nikitin,
       Zh. Fiz. Khim., 36:2277 (1962).
155.   N. V. Akshinskaya, A. V. Kiselev, and Yu. S. Nikitin, Zh. Fiz. Khim.,
       37:927 (1963); 38:488 (1964).
156.   A. V. Kiselev and Ya. I. Yashin, Neftekhimiya, 4:494 (1964).
157.   A. V. Kiselev, Yu. S. Nikitin, R. S. Petrova, K. D. Shcherbakova, and Ya. I.
       Yashin, Anal. Chem., 36:1526 (1964).
158.   N. V. Akshinskaya, A. V. Kiselev, Yu. S. Nikitin, R. S. Petrova, V. K.
       Chuikina, and K. D. Shcherbakova, Zh. Fiz. Khim., 36:1121 (1962).
159.   I. Yu. Babkin and A. V. Kiselev, Zh. Fiz. Khim., 36:2448 (1962).
160.   A. I. Bruk, D. A. Vyakhirev, A. V. Kiselev, Yu. S. Nikitin, and N. M.
       Olif'erenko, Neftekhimiya, 7:145 (1967).
161.   S. P. Zhdanov, in collection: Methods of Structure Examination for Finely
       Divided and Porous Bodis, Moscow, Izd. Akad. Nauk SSSR, 1958, p. 117.
162.   S. P. Zhdanov, A. V. Kiselev, and Ya. I. Yashin, Neftekhimiya, 3:417
       (1963).
163.   R. D. Schwartz, D. J. Brasseaux, and R. G. Mathews, Anal. Chem., 38:303
       (1966).
164.   J. J. Kirkland, in collection: Gas Chromatography 1964, A. Goldup (ed.),
       London, Institute of Petroleum, 1965, p. 285.
165.   L. H. Klemm and S. K. Airee, J. Chromatog., 13:40 (1964).
166.   A. V. Kiselev, in collection: Gas Chromatography, Transactions of the
       First All-Union Conference on Gas Chromatography, Moscow, Izd. Akad.
       Nauk SSSR, 1960, p. 45.
167.   A. V. Kiselev and K. D. Shcherbakova, in collection: Gas Chromatographie
       1959, R. Kaiser and A. G. Struppe (eds.), Berlin, Akademie Verlag, 1959,
       p. 198.
168.   A. V. Kiselev, in collection: Gas Chromatography 1962, M. van Swaay
       (ed.), London, 1962, p. 34.
169.   A. V. Kiselev, Zh. Fiz. Khim., 35:233 (1961).
170.   A. V. Kiselev, in collection: Gas Chromatography 1964, A. Goldup (ed.),
       London, Institute of Petroleum, 1965, p. 238.
171.   V. S. Vasil'eva, A. V. Kiselev, Yu. S. Nikitin, R. S. Petrova, and K. D.
       Shcherbakova, Zh. Fiz. Khim., 35:1386 (1961).
172.   A. V. Kiselev, E. A. Paskonova, R. S. Petrova, and K. D. Shcherbakova,
       Zh. Fiz. Khim., 38:161 (1964).
173.   R. L. Gale and R. A. Beebe, J. Phys. Chem., 68:555 (1964).
174.   L. D. Belyakova, A. V. Kiselev, and N. V. Kovaleva, Bull. Soc. Chim.
       France, No. 1:285 (1967).
175.   I. Halasz and C. Horwath, Nature, 197:71 (1963).
176.   C. G. Pope, Anal. Chem., 35:654 (1963).
177.   I. Halasz and C. Horwath, Anal. Chem., 36:2226 (1964).
178.   E. Cremer, Angew. Chem., 71:512 (1959).
179.   A. V. Kiselev and Ya. I. Yashin, Zh. Fiz. Khim., 40:603 (1966).
180.   A. V. Kiselev and Ya. I. Yashin, in collection: Gas Chromatographie 1965,
       Berlin, Akademie Verlag, 1965, p. 43.

181.   A. V. Kiselev and Ya. I. Yashin, Zh. Fiz. Khim., 40:429 (1966).

182.   L. C. Case, J. Chromatog., 6:381 (1961).

183.   J. van Rysselberge and M. von Stricht, Nature, 193:1281 (1962).

184.   A. V. Kiselev, G. M. Petov, and K. D. Shcherbakova, Zh. Fiz. Khim.,
        41:1418 (1967).

185.   H. P. Burchfield and E. E. Storrs, Biochemical Applications of Gas Chro-
        matography [Russian translation], Moscow, Izd. Mir, 1964 [English edition:
        New York, Academic Press].

186.   A. V. Kiselev, I. L. Migunova, and Ya. I. Yashin, in collection: Gas Chro-
        matography, Moscow, Izd. NIITÉKhIM, Issue 6, 1967, p. 84.

187.   D. White, Nature, 179:1075 (1957).

188.   D. White and C. T. Cowan, in collection: Gas Chromatography, Transactions
        of the Second International Symposium on Gas Chromatography in Amsterdam
        [Russian translation], Moscow, Izd. In. Lit., 1964, p. 112.

189.   M. A. Hughes, D. White, and A. L. Roberts, Nature, 184:1796 (1959).

190.   J. V. Mortimer and P. L. Gent, Nature, 197:789 (1963).

191.   C. F. Spencer, Anal. Chem., 35:592 (1963).

192.   J. V. Mortimer and P. L. Gent, Anal. Chem., 36:754 (1964).

193.   V. von Sticht and J. von Rysselberge, J. Gas Chromatog., 1:29 (1963).

194.   B. Versino, F. Geiss, and G. Barbero, Z. Anal. Chem., 201:20 (1964).

195.   E. W. Cieplinski, Anal. Chem., 37:1160 (1965).

196.   J. A. Favre and L. R. Kallenbach, Anal. Chem., 36:63 (1964).

197.   L. B. Rogers and A. G. Altenau, Anal. Chem., 35:915 (1963).

198.   A. G. Altenau and L. B. Rogers, Anal. Chem., 36:1726 (1964).

199.   A. G. Altenau and L. B. Rogers, Anal. Chem., 37:1432 (1965).

200.   H. Kelker, Ber. Bunsen Gesellschaft für Physikal. Chemie, 67:698 (1963).

201.   H. Kelker, Z. Anal. Chem., B198:254 (1963).

202.   H. Kelker, in collection: Gas Chromatographie 1965, H. P. Angele and
        H. G. Struppe (eds.), Berlin, Akademie Verlag, 1965, p. 271.

203.   F. M. Wachi, Dissertation, Abstr., 20:53 (1959).

204.   R. S. Juvet and F. M. Wachi, Anal. Chem., 32:290 (1960).

205.   W. W.Hanneman, C. F. Spencer, and J. F. Johnson, Anal. Chem., 32:1386
        (1960).

206.   W. W. Hanneman, J. Gas Chromatog., 1:18 (1963).

207.   C. G. Scott and C. S. G. Phillips, in collection: Gas Chromatography 1964,
        A. Goldup (ed.), London, Institute of Petroleum, 1965, p. 266.

208.   C. G. Scott, in collection: Gas Chromatography 1962, M. van Swaay (ed.),
        London, 1962, p. 46.

209.   G. Urbach, Anal. Chem., 36:2368 (1964).

210.   O. L. Hollis, Anal. Chem., 38:309 (1966).

211.   N. A. Malafeev, I. P. Yuidina, and N. M. Zhavoronkov, Usp. Khim., 31:710
        (1962).

212.   F. Dupire, Z. Anal. Chem., 170:317 (1959).

213.   E. Adlard and B. Whitham, in collection: Gas Chromatography, Trans-
        actions of the Second International Symposium on Gas Chromatography in
        Amsterdam [Russian translation], Moscow, Izd. In. Lit., 1961, p. 325.

214. P. Perrero, Ind. Chim. Belge, 25:237 (1960).
215. O. H. Orr and J. E. Callen, J. Am. Chem. Soc., 80:249 (1958).
216. J. S. Lewis and H. W. Patton, in collection: Gas Chromatography, V. J. Coates, et al. (eds.), New York, Academic Press, 1958, p. 145.
217. J. Zulaica and G. Guiochon, Compt. Rend., 225:524 (1962).
218. J. L. Ogilvic, M. C. Simmons, and G. P. Hinds, Anal. Chem., 30:25 (1958).
219. N. N. Nicolonides, J. Chromatog., 4:496 (1960).
220. F. E. de Boer, Nature, 185:915 (1960).
221. J. Anderson, in collection: Progress in Industrial Gas Chromatography, H. A. Szymanski (ed.), Plenum Press, New York, 1961, p. 73.
222. A. V. Kiselev, Yu. S. Nikitin, N. K. Savinova, I. M. Savinov, and Ya. I. Yashin, Zh. Fiz. Khim., 38:2328 (1964).
223. R. W. Solomon, Anal. Chem., 36:476 (1964).
224. O. T. Chortyk, W. S. Scholtzhauer, and R. L. Stedman, J. Gas Chromatog., 3:394 (1965).
225. A. V. Kiselev, I. A. Migunova, F. Onushka, K. D. Shcherbakova, and Ya. I. Yashin, in collection: Gas Chromatography, Moscow, Izd. NIITÉKhIM, Issue 6, 1967, p. 75.
226. K. D. Nenicescu, Organic Chemistry [Russian translation], Vol. 2, Moscow, IL, 1963.
227. C. Karr and J. R. Comberioti, J. Chromatog., 18:394 (1965).
228. R. S. Gohlke, Anal. Chem., 31:535 (1959).
229. C. H. Deal, J. W. Otvos, V. H. Smith, and P. S. Zucco, Anal. Chem., 28: 1958 (1956).
230. A. Liberti, L. Conti, and V. Crescenzi, Nature, 178:1067 (1956).
231. A. L. Liberti, G. Costa, and E. Pauluzzi, Chim. Ind., 38:674 (1956).
232. F. M. Noble, K. Abel, and P. W. Cook, Anal. Chem., 36:1421 (1964).
233. J. E. Lovelock and S. R. Lipsky, J. Am. Chem. Soc., 82:431 (1960).
234. A. J. P. Martin and A. T. James, Biochem. J., 63:138 (1956).
235. J. T. Walsh and C. Merritt, Anal. Chem., 32:1378 (1960).
236. J. Janak and J. Novak, Collection Czech. Chem. Commun., 24:384 (1959).
237. N. H. Roy, Analyst, 80:863 (1955).
238. F. Cacace, R. Cipollini, and G. Perez, Science, 132:1253 (1960).
239. W. H. T. Davison, S. Slaney, and A. L. Wragg, Chem. Ind., 1954, p. 1356.
240. J. Janak, Nature, 185:684 (1960).
241. J. Janak, Collection Czech. Chem. Commun., 25:1780 (1960).
242. A. T. James and A. J. Martin, Biochem. J., 50:679 (1952).
243. V. L. Keibal, A. V. Kiselev, I. M. Savinov, V. L. Khudyakov, and Ya. I. Yashin, Zh. Fiz. Khim., 41:2234 (1967).
244. A. Wehrli and E. Kovats, Helv. Chem. Acta, 42:2709 (1959); L. S. Ettre, Anal. Chem., 36:31A (1964).
245. J. Janak, J. Chromatog., 3:308 (1960).
246. M. Shinglar, Practical Gas Chromatography [Russian translation], Moscow, Izd. Khimiya, 1964, p. 56.
247. E. Bayer, Gas Chromatography [Russian translation], Moscow, IL, 1961 [English edition: New York, American Elsevier].

248. J. H. Purnell, Gas Chromatography, New York, John Wiley, 1962, p. 397.

249. N. V. Akshinskaya, V. E. Beznogova, A. V. Kiselev, and Yu. S. Nikitin, Zh. Fiz. Khim., 36:2277 (1962).

250. J. C. Giddings, Anal. Chem., 35:1338 (1963).

251. A. V. Kiselev, Discussions Faraday Soc., 40:205 (1965).

252. J. A. Walker, Nature, 209:197 (1966).

253. E. Zielinski, Chem. Anal., 11:67 (1966).

254. R. L. Hoffman, G. R. List, and C. D. Evans, Nature, 211:965 (1966).

255. G. A. Drennan and R. A. Matuld, J. Chromatog., 34:77 (1968).

256. T. N. Gvozdovich, A. V. Kiselev, and Ya. I. Yashin, Neftekhimiya, 8:476 (1968).

257. W. F. Wilhite and O. L. Hollis, J. Gas Chromatog., 6:84 (1968).

258. N. K. Bebris, G. E. Zaitseva, A. V. Kiselev, Yu. S. Nikitin, and Ya. I. Yashin, Neftekhimiya, 8:481 (1968).

259. Yu. S. Nikitin, in collection: Proceedings of the Sixth Symposium on Gas Chromatography, Berlin, 1968; J. Chromatog., (in press).

260. A. V. Kiselev and Ya. I. Yashin, in collection: Gas Chromatography, A. B. Littlewood (ed.), London, 1968.

261. G. C. Goretti, A. Liberti, and G. Nota, J. Chromatog., 34:96 (1968).

262. F. Geiss, B. Versino, and H. Schlitt, Chromatographia, 2:9 (1968).

263. A. V. Kiselev, in collection: Proceedings of the Sixth Symposium on Gas Chromatography, Berlin, 1968; J. Chromatog., (in press).

264. F. Onuska, J. Janak, K. Tesarik, and A. V. Kiselev, J. Chromatog., 34:81 (1968).

265. S. T. Sie and G. W. A. Rijnders, Anal. Chim. Acta, 38:21 (1967).

266. S. T. Sie and G. W. A. Rijnders, Separation Science, 1968 (in press).

267. S. T. Sie, J. P. A. Bleumer, and G. W. A. Rijnders, in collection: Preprints of the Seventh International Symposium on Gas Chromatography, Copenhagen, June 25-28, 1968, C. L. Harbourn (ed.), London, Institute of Petroleum, 1968, paper 14.

Chapter VI

Practical Techniques

PREPARATION OF ADSORBENTS

FOR GAS CHROMATOGRAPHY

Graphitized Carbon Blacks

These are prepared by heating ordinary carbon blacks to about 3000°C under vacuum, in an inert gas, or in a reducing atmosphere [1, 2]. Volatile tarry substances are removed below 1000°C, while above 1000°C the crystallites grow, side chains and oxides being destroyed and the specific surface decreasing. Thermal carbon blacks of low s at 2200-3200°C tend to acquire completely parallel crystallites; at 3000°C the particles become polyhedra (Fig. 1) with faces formed by graphite crystals growing from within the particles [1, 2]. For thermal carbon blacks with particularly large particles, this treatment causes most of the surface of the polyhedra to consist of basal faces of graphite, while the effects of inhomogeneous sites at edges and corners become very small [3]. Hence the behavior of a thermal carbon black graphitized at 3000°C is very similar to that of a uniform basal face of graphite.

A graphitized carbon black has no ions, functional groups, or π-bonds at the surface [4]. The adsorption is due mainly to attraction under dispersion forces [4], so a graphitized carbon black is a type I (nonspecific) adsorbent (pp. 11 and 16-40 of Chapter II).

Treatment of channel carbon black at 3000°C produces a Graphon-type material of higher s (~90 m^2/g) with less uniformity.

Acetylene carbon blacks are heated to ~2500°C in the main process of production, so they can be used in chromatography without additional treatment at 3000°C; s is ~60 m^2/g.

Particles of carbon black can clump together into more or less branched chains (carbon-black structures) [5]. This facilitates the use of the fine powders. Prolonged working readily produces fairly large clumps, which can then be sieved to yield the desired fraction; but the clumps will not stand much mechanical stress, so particular care is needed in filling columns.

We have seen in Chapter V that graphitized carbon black can be mounted on modified macroporous silica gel [6] or porous Teflon (the black enters the wide pores [7], the materials acting as inert supports). The particles may also be melted into the surfaces of nonporous polyethylene [8] or glass [9] spheres, though the thermal stability is poor in the first case.

Graphitized carbon black may also be mounted on the inner surface of a capillary [10]. The uptake of water is slight, so there is no need to fire the black before filling the column.

Teflon and Other Porous Polymers

Teflon (polytetrafluoroethylene), $(-CF_2-CF_2-)$, is a carbon-chain crystalline polymer of molecular weight 500,000 to 2,000,000. Its relatively high thermal stability gives it advantages over many other organic polymers in gas chromatography.

The crystal structure is disrupted at 327°C, and the material becomes transparent and highly elastic, this condition persisting up to the decomposition point (~415°C) [11].

Teflon is produced by polymerization of tetrafluoroethylene in the presence of peroxide catalysts. The more ordered crystalline nuclei are usually accompanied by fibrous amorphous parts. Porous Teflon may have s up to 10 m^2/g. Argon is [12] adsorbed more strongly by Teflon than is nitrogen, although the latter has a large quadrupole moment; Teflon is therefore a nonspecific adsorbent, whose surface is even less specific than that of graphitized thermal carbon black.

Adsorbed molecules of n-hexane lie mainly between the chains; this gives a high energy of dispersion interaction and hence high Q_0 with low entropy. Q for benzene is less than that for n-hexane [13].

It is difficult to pack a column uniformly with Teflon, because the powder does not pour well and the particles tend to stick together; special grades of granulated Teflon are needed.

Other porous polymers are also used in gas chromatography.
Porous nonpolar polymers may be produced, for instance, by block
polymerization of styrene or t-butylstyrene with divinylbenzene as
cross-linking agent [14], e.g., in a mixture of 21.8% styrene, 18.2%
technical (55%) divinylbenzene, and 60% diluent (toluene and dode-
cane). The polymer is produced as spheres by suspension polymerization. The s for styrene-base porous polymers may be 100 m^2/g
or more. These materials show nonspecific adsorption, and they
have N_2 running ahead of O_2, and H_2O ahead of methanol and vinyl
chloride. Porous polyaromatic polymers give good separation for
a mixture of water with formic, acetic, and propionic acids, all of
which give symmetrical peaks. Stable macroporous polymers are
produced by solvent sublimation (p. 72, Chapter III).

Salts

Nonporous inorganic salts and hydroxides are usually supported on macroporous solid supports, especially diatomite [15],
silica gel [16], and alumina gel [17]. The salt may be deposited on
the support as follows [15]. The salt (20-25% by weight of the support) is dissolved in distilled water, and the support is mixed into
the solution. The water is evaporated off with continuous stirring,
and then the material is heated to the melting point of the salt. This
produces uniform distribution of the salt over the support. High-
temperature analyses of polyphenyls (p. 205, Chapter V) have [18]
been performed with Chromosorb P bearing LiCl, $CaCl_2$, CsCl, and
other salts.

Porous complex salts are produced [19] by partial removal of
the coordinated compound on heating; the latter may be ammonia,
water, pyridine, bipyridyl, 1,10-phenanthroline, and various other
compounds. The best thermal stability occurs in complexes of copper salts with bipyridyl and 1,10-phenanthroline [20].

The complex salt is prepared by adding 0.01 mole of compound
to a solution of 0.01 mole of the copper salt, the mixture being heated
until the reaction is complete. The quantity of water needed for
$Cu(NO_3)_2$ is 25 ml, whereas $CuSO_4$ and $CuCl_2$ need about 250 ml. The
solution is slowly cooled to room temperature, and crystals of the
complex are deposited. The dried crystals are placed in the column
and treated with inert gas at 150-200°C to remove part of the com-
plexing agent. The products are $Cu(Phen)_2(NO_3)_2$, $Cu(Bipy)(NO_3)_2$,
$Cu(Phen)Cl_2$, and $Cu(Phen)SO_4 \cdot H_2O$ (here Bipy denotes bipyridyl and

Phen denotes 1,10-phenanthroline). The most stable is $Cu(Phen)SO_4 \cdot$ $\cdot H_2O$, which does not decompose below 500°C [20]. These porous complex salts may be used with program temperature control.

Salts provide selectivity for many liquid compounds; a surface with the appropriate chemical nature is produced by choice of the salt, and high selectivity can be combined with high performance.

Zeolites Types A and X

These are used without binding agents (as pressed discs [21] or clumps of crystals [22]) or bonded discs (clay being the usual bonding agent).

The basic unit in the zeolite lattice is a cubo-octahedron composed of 24 atoms of Al and Si with 48 oxygen atoms. These figures in type A zeolites are linked into a simple cubic lattice, while in type X the structure is more open.

Zeolites act as molecular sieves and also differ considerably from other adsorbents. The uniform pore size means that they take up only molecules that can penetrate into the pores. The force fields are superimposed in the narrow pores, so the capacity is high even at elevated temperatures. The SiO_4 and AlO_4^- tetrahedra are linked in a framework bearing exchangeable cations (p. 49, Chapter II), so they have high affinity for molecules with peripherally localized electron density (π-bonds, nonbonding pairs on N or O; see pp. 13-14, Chapter II). Zeolites are therefore used in purifying and drying gases [20] in many technical applications as well as in gas chromatography.

Zeolites are used in gas chromatography mainly with substances of low boiling point. However, water and CO_2 have marked effects on the V_R, so these must be carefully removed from samples and carrier gases. Careful purification allows [25] zeolites to be used in automatic analysis for prolonged periods; drying with P_2O_5 allowed a 400×0.4 cm column of NaX (13 X) to work for more than 10,000 cycles without change in characteristics although the initial gases contained water.

The separation is dependent on the cationic form [26]; type AgX produces irreversible absorption of H_2, CO, and unsaturated hydrocarbons, which produce reduced silver [27].

Table 13. Approximate Channel Sizes for Some Zeolites

Zeolite	Size, Å	Zeolite	Size, Å
NaA (4A)	~4	NaX (13X)	10
CaNaA (5A)	4-5	CaNaX (10X)	8-10

The separating power is also much dependent on the amount and composition of the bonding agent, the completeness of cation exchange, the crystal size, and the synthesis conditions. Large crystals (15 μ or more) produce worse separations than do small ones (2-3 μ) [28].

The performance of a zeolite column may be increased by a factor 6 or so if the small crystals (e.g., of CaA) are supported on a Chromasorb [29] instead of being mixed with a binder; this shortens the diffusion paths and makes all the crystals readily accessible.

High performance is also attained by coating the inside of a capillary with the small zeolite crystals [30].

Molecules that cannot enter the pores may still be adsorbed on the external surface (p. 50) and also in the secondary pores produced by the macroporous binder.

Type CaA has been used [31] to separate the heavier normal hydrocarbons (which are almost irreversibly adsorbed in a certain temperature range) from aromatic and branched-chain ones, which do not penetrate into the pores and so pass through the layer of zeolite. For instance, CaA has been used for industrial separation and determination of normal hydrocarbons in benzene fractions [32].

Zeolites before use must be baked at 450-500°C for 4-5 h (the crystal structure is altered at higher temperatures), especially if they are to be used for separation of $O_2 - N_2$ mixtures. The working time and temperature needed with light hydrocarbons and other compounds may be reduced by the use of zeolites bearing water [33].

Zeolites prepared for gas chromatography must be stored in sealed vessels.

Table 13 gives approximate channel sizes for some zeolites.

Silica Gels and Porous Glasses

Silica Gels are produced from orthosilicic acid, which is
made by hydrolysis of the chloride or by reaction of mineral acids
with waterglass. The condensation occurs by extension, branching,
and ring closure in the Si—O chains; the solution then contains
macromolecules, which increase the viscosity and alter the optical
properties (silica sol). The sol SiO_2 particles take the shape of
minimum surface (spherical) [34, 35], whose size depends on the
method of preparation and ranges from 20-30 Å to 150 Å, or rather
larger with special methods of preparation [36]. The viscosity of
a concentrated silica sol increases with time on account of particle
growth and aggregation (gel formation). Drying produces a porous
solid of globular structure (xerogel). To produce a silica gel of specified structure, the hydrogel before drying may be treated with solutions of salts, dehydrating agents [37], or surface-active materials [38]. Reduction in surface tension causes the packing of the
globules in the xerogel to become more open (larger pores) and vice
versa. The structure of the xerogel pores is also affected by the
pH of the sol or hydrogel; for instance, washing at pH < 4 tends to
produce small pores.

However, none of these methods of structure control gives a
gel whose pores are large enough for use in gas chromatography of
liquids and solids.

Hydrothermal treatment (high temperatures in the presence
of water vapor, see Chapter III, p. 70, and also [39, 40]) is used to
adjust the structure further, the effects being dependent on the initial state of the gel, the chemical composition, the initial porosity,
the temperature used, and the steam pressure.

The surface and volume of the pores are reduced to a greater
extent when the pore size is small [41, 42], so a specimen with a
large range in pore size shows a greater reduction in s and increase
in pore diameter [41, 42], because the small pores are the first to
be lost.

Hydrothermal treatment at up to 700°C will give macroporous
silica gel and other xerogels with very small s and with frameworks
formed of fairly regularly packed large globules similar in size.
The large pores are the gaps between globules; they are reasonably
uniform and are accessible to molecules of a great variety of sizes

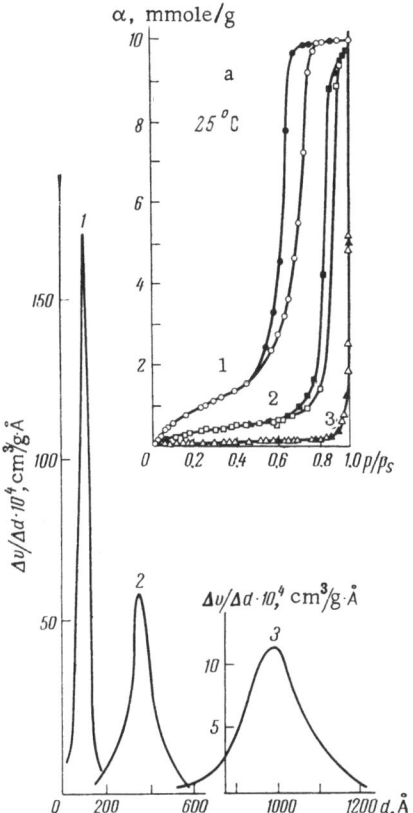

Fig. 181. a) Adsorption and desorption isotherms for benzene on several hydrothermally
treated silica gels: 1) s = 300 m^2/g, 2) s = 117
m^2/g, 3) s = 23 m^2/g; b) corresponding pore-
size distributions.

and structures. However, even more rigorous treatment (Fig. 38
in Chapter III) converts the globular structure to a spongy one, with
onset of local crystallization. The volume and accessibility of the
macropores are reduced, and the gel becomes unsuitable as an adsorbent or support for gas chromatography. The best support properties occur in the most homogeneous macroporous silica gels, with
s of 5-10 m^2/g (p. 70, Chapter III). Figure 181a shows adsorption
and desorption isotherms for benzene on several hydrothermally
treated silica gels, while Fig. 181b shows the corresponding distributions of pore sizes [43].

Table 14. Characteristics of Some Soviet
Silica Gels for Gas Chromatography

Gel	s, m^2/g	Pore vol., cm^3/g	Mean pore size, Å
Microporous			
KSM	700	0.35	20
ShSM	900	0.25	10
Intermediate pores			
ShSk	300	0.90	120
MSK	210	0.80	150
KSK	350	1.08	120
Macroporous			
MSA-2	80	0.75	380
MSA-1	30	0.55	750

All commercial silica gels made from $SiCl_4$ or waterglass
(see above) have medium to fine pores, with broad distributions from
200 Å downwards; they are thus suitable only for separating inert
gases and light hydrocarbons. Larger pore sizes and more uniform
geometry are needed for gas-chromatographic separation of liquids.

Chapter V gives examples of the use of silica gels with various
s and various pore sizes; it also gives literature citations. Table 14
gives the characteristics of some Soviet silica gels suitable for gas
chromatography. Aerosilogels are most homogeneous [77].

Porous glasses have latterly come into use as adsorbents
[44, 45]. Grebenshchikov [46] showed that alkali borosilicate glasses,
when given appropriate heat treatment, become chemically unstable
in acids and alkalis. Acid treatment gives porous glasses by selective attack, only alkali-metal oxides and B_2O_3 dissolving, and the
SiO_2 remaining in porous form or as a silica gel deposit in pores
[47, 48].

The structure and characteristics of this porous glass are dependent on the leaching conditions and also on the composition and
heat treatment of the initial glass, which determine the distribution
of the B_2O_3 and alkali-metal oxides in the silica framework [48].
This makes it possible to produce specimens with preset pore sizes
from 8 to 1000 Å with only narrow spreads [48], which are of great
interest in gas chromatography.

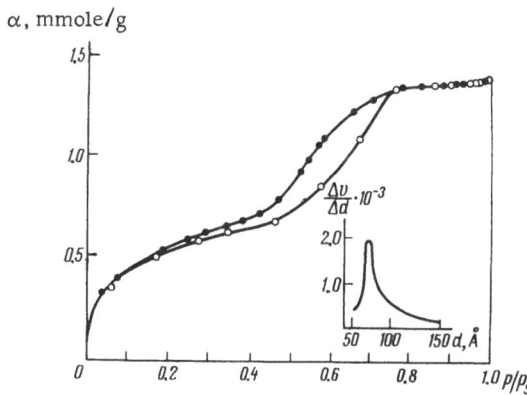

Fig. 182. Adsorption of benzene on fine-pore glass and pore-size distribution deduced from the desorption branch.

Zhdanov [49] considers that porous glasses are formed without rupture of Si—O— Si bonds, so the silica framework is a residue formed by the silicon—oxygen tetrahedra in the initial glass; the initial and final structures are related. The initial glass contains areas of chemical inhomogeneity of considerable size (100 Å or more), and treatment at 500-700°C can produce profound changes in the submicrostructure.

The B_2O_3 does not combine with the SiO_2 in these soda borosilicate glasses; the B and Si atoms do not form a common network. The anomalous behavior of these glasses occurs partly because the state of coordination of the boron is altered, though this is not the only structural change. Small amounts of Na_2O in a $B_2O_3-SiO_2$ glass give rise to large regions containing B_2O_3 and Na_2O, which grow on heating and produce borate complexes, in which the polar $Na^+[B(O_{1/2})_4]^-$ groups are surrounded by neutral $B(O_{1/2})_3$ ones, the association of these groups increasing the regions of inhomogeneity.

These regions of inhomogeneity determine the detailed structure of the silica left after acid treatment. Zhdanov [48] showed that alkali largely disrupts the detailed structure formed by the finely divided silicic acid within the coarser structure, the last corresponding to the part of the initial glass composed solely of silica, which remains almost unaltered by acids. Alkali increases the total pore volume and reduces s substantially. Adsorption tests and electron micrographs [48] show that pores of radius 300-500 Å predominate.

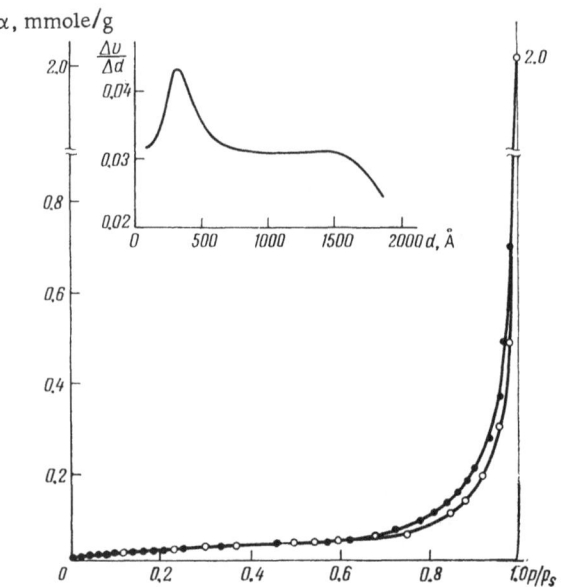

Fig. 183. Adsorption of benzene on coarse-pore glass and pore-size distribution deduced from the desorption branch.

The initial glass is usually DV-1, whole composition is 7% Na_2O, 23% B_2O_3, 70% SiO_2. The required porosity after leaching is provided by heat treatment at 500-700°C, which is followed by crushing to the grain size required for the columns and separation of the 0.25-0.5 mm fraction (or, better, the 0.2-0.25 mm fraction). These grains are treated with 3 N HCl at 50°C with continuous stirring; the product is washed free from chloride and is dried to constant weight at 150-200°C.

Firing of the initial glass at 550°C for 20 h produces a pore size of 30-60 Å.

Figure 182 shows the adsorption isotherm for benzene and the fairly narrow pore-size distribution for such a glass; the mean pore diameter is ~60 Å.

Adjustment of the time of treatment with HCl gives control of the pore depth; bulk-porous and surface-porous materials can be produced [50], the latter being especially convenient (Fig. 47). The mean pore diameter is almost independent of the time spent in the HCl.

Table 15. Heats of Adsorption (kcal/mole)
for Aromatic Hydrocarbons on Hydroxylated
Silica Adsorbents at Degrees of Filling <0.001

Adsorbent	Hydrocarbon			
	Benzene	Toluene	Ethylbenzene	Isopropylbenzene
Macroporous silica gal....	9.5	11.5	13.7	14.6
Macroporous galss.......	9.8	11.5	13.5	14.1

Larger pores arc produced by treating the acid-etched glass with 0.5 N KOH at 20°C for 1-2 h, which dissolves the finer silica particles, greatly reduces s, and increases the pore volume [48].

It is also possible to produce unusual bidisperse systems having two sets of pores differing considerably in size [51]. In particular, it is of value to have a combination of large pores (which act mainly as transport channels) with smaller but homogeneous ones, which produce most of the adsorption. However, this is a topic for further study.

Figure 183 shows an adsorption isotherm for benzene for such a glass, and also the pore-size distribution, which is dominated by two groups, one with a mean size of ~250 Å and the other covering the broad region 500-1500 Å.

The structural control makes these glasses very useful in gas chromatography. Molecular sieve glasses (pore size 10 Å) can be used as are zeolites for separating low-boiling gases. They have higher mechanical strength than zeolites and are completely stable to acids.

Pore sizes of 30 to 100 Å can be used in the analysis of gases and light hydrocarbons; the uniformity gives porous glasses advantages over silica and alumina gels in this respect [44]. The high separating power for such gases (e.g., CH_4, C_2H_6, and C_2H_4) persists even at 95°C [44]. The uniformity is greater than that of silica gel, which means that trace components in very pure substances can be detected, e.g., in monomers [44].

Glasses with pore sizes over 500 Å can be used to separate liquid mixtures of normal and aromatic hydrocarbons.

The surface of a thoroughly washed porous glass is the same in nature as that of silica gel. Table 15 compares the Q for aromatic hydrocarbons on macroporous silica gel and macroporous glass ($s \approx 30$ m^2/g) at similar degrees of filling [52]; the values, which are sensitive to the degree of hydroxylation of the surface (p. 44, Chapter II), agree well.

Porous glasses must be prepared as for silica gels for use in gas chromatography.

Activated Charcoal

This is widely used as an adsorbent that takes up various substances very strongly and mainly unspecifically. Such charcoals are usually prepared by removal of tarry matter from the raw charcoal and treatment with H_2O and CO_2 at high temperatures, or by treatment of organic materials with salts (K_2S, $ZnCl_2$) followed by firing in the absence of air and washing the product with water. Very uniform fine-pore activated charcoals may be made by pyrolysis of polyvinylidene chloride (Saran charcoal).

The framework consists of open and irregularly packed sets of grids of 6-membered carbon rings; these sets form small contacting crystallites, but the structure is much less ordered than in graphite. The particles are of very small size, often 23×9 Å. The edges of the carbon grids bear various groups, including $C=O$; the composition of these is very much dependent on the method of preparation.

The pore structure is complex. Dubinin [53] divides the pores into three types: strongly adsorbing micropores [10-30 Å], transition pores (where capillary condensation occurs), and macropores; on this basis he [54] divides charcoals into two limiting structural types. Those of the first type contain mainly micropores (size ~10 Å) to a volume of 0.4-0.5 cm^3/g, while the second type contains micropores and also larger pores (sometimes with the latter predominating) with radii over 30 Å. Here s is determined by the micropores and is 400 to 900 m^2/g in accordance with the formal definition of the BET method.

The surfaces are geometrically and chemically very inhomogeneous; they may be oxidized to various extents and so take up

water, though the uptake of water is not large if the surface is not
too heavily oxidized.

Activated charcoal has been used in gas chromatography (see
pp. 152, 153, and 160 of Chapter V) with low-boiling gases (in particular, inert gases) and volatile hydrocarbons. Microporous active
charcoals suitable for this purpose include Saran charcoals (with
only a small volume of the smallest and uniform pores); less suitable are SKT-2 (made from peat by the K_2S method) and AG (made
from coal).

PREPARATION OF CAPILLARY
ADSORPTION COLUMNS

Preparation of Glass Capillaries

Glass capillaries are sometimes the most suitable for capillary columns, and they do not give rise to the difficulties due to
strong adsorption by the wall found with metal capillaries. Glass
capillaries are fairly inert, and they do not produce catalytic reactions at working column temperatures. The transparency allows
one to check on the uniformity of distribution of stationary phases.
Finally, they can also be made directly in the laboratory.

Desty et al. [55] described a device for drawing glass capillaries; Kalmanovski et al. [56] have described a similar device (Fig.
184), which gives a long capillary wound into a compact spiral.

The glass tube 1 (usually soft glass) of outside diameter ~6
mm and inside diameter ~2 mm is fed by the rollers 2 at about 1
cm/min into the cylindrical oven 3 (internal diameter ~10 mm) made
of nichrome strip (2.5×0.5 mm). This oven is supplied with a current of up to 60 A at 5 V, which produces a temperature sufficient
to soften the glass.

A thin capillary is drawn from the end of the heated part on
account of the difference in speed between the feed rollers 2 and
pulling rollers 4. The size can be adjusted via the ratio of these
speeds.

The straight capillary passes into the stainless-steel tube 5,
which is electrically heated (~100 A at 2 V) to a temperature below

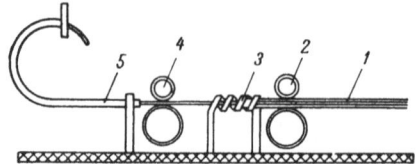

Fig. 184. Apparatus for making glass capillaries:
1) glass tube; 2) feed rollers; 3) nichrome oven;
4) pulling rollers; 5) tube for winding capillary
into a spiral.

the melting point of the glass, but sufficient to soften it. The capillary emerges as a spiral (radius is governed by that of the tube) and is taken up by a rotating drum.

A piece of initial tube ~140 cm long will give a 50 m capillary of outside diameter ~1 mm and bore 0.3-0.4 mm. Another piece of the initial tube may be sealed onto the first in order to produce a longer capillary. Existing spirals may also be joined together by making the ends to be joined as smooth as possible (notches are best made on the inside of the spiral). The two ends are pressed firmly together and are inserted in a small gas flame; the diameter of the end of the burner should not exceed 2 mm, e.g., a pulled-down piece of glass tube. The heated ends are joined and pulled apart slightly. The variation in internal diameter at the joint need not exceed 2%.

A more inert internal surface may be produced by chemical modification during drawing [57]; the initial tube is filled with trimethylchlorosilane and one end is sealed. The tube is drawn into a capillary from the open end.

Production of a Porous Adsorbing Layer

on the Inner Surface of a Capillary Column

A porous layer may be produced on a glass capillary either by etching the surface itself or by deposition of an adsorbent.

In the first case, the tube is made of sodium borosilicate glass [58], which is drawn as in Desty's method [55] and then is treated (20 h at 550°C) to produce the appropriate structure [58], which is followed by etching in 0.1 N HCl at 25°C for 5 min and washing with water. The porous film may be up to 0.1 mm thick [58]. The C_2-C_4 hydrocarbons have [58] been separated on such a capillary of length 10 m and internal diameter 0.5 mm.

Prolonged treatment with NH_4OH will produce a porous layer on soft glass [59]; such a column 80 m × 0.27 mm has provided complete separation of hydrogen isotopes and isomers (Fig. 86).

Ordinary soft glass may also be etched with 20% NaOH in water [60, 61], preferably at 100°C for 6 h, with the NaOH passing slowly through the column (10 ml/h). The capillary is then washed with water until the reaction is neutral, followed by alcohol and ether; the final drying is with dry nitrogen. Such columns have been used in the separation of deuterated compounds [62] (Fig. 99).

Deposition of an Adsorbing Layer
in a Capillary Column

The material is deposited from a suspension or sol. For instance, a silver-plated copper column has been coated with carbon black graphitized at 3000°C by passage of a suspension in a liquid of high density (e.g., CF_3CCl_3, methylene bromide) [10]. The layer is carefully dried by gradual heating from the open end. Such a graphite column may be of very high performance (Fig. 133).

A capillary adsorption column may also be made by passage of a silicic acid sol in a water—isopropanol mixture, with subsequent drying in a current of dry argon [63]. Column of nylon, stainless steel, and copper have been used.

A column coated with silica gives excellent separation of aromatic hydrocarbons [64]. The capacity of a glass capillary column may be increased by deposition of silica gel (prepared as in [40]) by repeated passage of the sol through the column.

Preparation of Packed Capillary Columns

Packed columns of small diameter (~1 mm) have certain advantages over ordinary packed columns [66], but there are difficulties in filling such columns uniformly. Halasz and Heine [65] described a method for glass capillaries based on Desty's apparatus [55]. A glass tube of internal diameter 2.2 mm is placed vertically with a carefully cleaned steel wire (1 mm) inside it. The annular space is filled with the adsorbent. Then the tube is placed horizontally in Desty's apparatus, the wire being carefully withdrawn as the tube is drawn. The adsorbent must have a grain size of 0.1-0.15 mm for this purpose.

Table 16. Residual Water Contents in Gases after Passage over Drying Agents

Drying agent	Flow rate, ml/min	Total volume, ml	Res. water, g/liter
$Mg(ClO_4)_2 \cdot 0.12H_2O$	216-228	1136-1200	0.2
$Mg(ClO_4)_2 \cdot 1.48H_2O$	115-233	1056-1258	1.5
Barium oxide	210-229	233-254	2.8
Alumina gel	213-233	251-275	2.9
P_2O_5	210-224	555-576	3.6
CaA zeolite	205-234	200-230	3.9
Mg perchlorate + $KMnO_4$	211-232	414-455	4.4
Li perchlorate	223-228	260-274	13
$CaCl_2 \cdot 0.18H_2O$	227-231	30-36	67
$CuSO_4 \cdot 0.02H_2O$	230-238	228-236	67
Silica gel	222-241	304-330	70
Ascarite	223-228	43-44	93
$CuSO_4 \cdot 0.21H_2O$	206-230	646-720	207
$NaOH \cdot 0.03H_2O$	221-230	17.4-18.2	513
Ba perchlorate	222-231	50-52	656
Magnesium oxide	220-222	22	753
$KOH \cdot 0.52H_2O$	222-229	18.2-18.6	939

Glass capillaries filled with alumina [65] and graphitized carbon black [66] have been made in this way and have been used to separate the isomeric $C_1 - C_5$ hydrocarbons (p. 167).

DRYING AND PURIFICATION
OF THE CARRIER GAS

Drying of the carrier gas is very important when specific adsorbents are used, since water vapor is almost irreversibly adsorbed by zeolites, silica and alumina gels, etc., which alters t_R and spoils the resolution. The changes in t_R are unacceptable for qualitative and quantitative purposes, especially in studies of adsorption isotherms and heats of adsorption, since (see p. 127, Chapter IV) Q and the uptake may be lower than those obtained with dry adsorbents under static conditions.

Trusell and Diehl [67] tested the performance of drying agents under dynamic conditions similar to those of gas chromatography. A flow of dry nitrogen was equilibrated with magnesium sulfate heptahydrate at 25°C, which gave a known water content; the nitrogen

Table 17. Absorbers for Gases and Organic Compounds

Substance	Absorber
Organic compounds	Microporous silica gel, activated charcoal, NaX (13 X) and CaX (10 X) zeolites
Normal hydrocarbons	CaA (5 A) zeolite*
Water	NaA (4 A) zeolite†
Unsaturated hydrocarbons	$HgSO_4 - H_2SO_4$, $Al_2O_3 + CuCl_2$
H_2S, NH_3, N_2O, NO_2, SO_2, CS_2, COS	NaX (13 X) and CaX (10 X) zeolites
H_2S and NH_3	KAD activated charcoal impregnated with $CuSO_4$
O_2	Charcoal impregnated with $K_2S_2O_8$, $KBrO_3$, and KIO_4
CO_2	Ascarite (90% NaOH) and calcium hydroxide (96%) + NaOH (4%)

* Only normal hydrocarbons are adsorbed; aromatic and branched-chain hydrocarbons do not penetrate into the pores of CaA zeolite.
† Only water, CO_2, and other compounds with small molecules are absorbed, other substances passing freely.

then passed through a tube containing the drying agent and thence to a trap in liquid nitrogen. The water collected in the trap was weighed. Table 16 gives results for 17 drying agents.

The most effective were magnesium perchlorate, barium oxide, alumina gel, phosphorus pentoxide, and CaA zeolite (molecular sieve 5 A); but the last is worse than the first by a factor 20.

Organic substances may be removed with microporous silica gel, activated charcoal, or NaX or CaX zeolite. Table 17 lists absorbers for various substances.

Filters of palladium − silver alloy [68] are used to purify the hydrogen used with ionization detectors; this alloy passes only hydrogen.

DETERMINATION OF EXIT TIME

OF AN UNADSORBED GAS

WITH AN IONIZATION DETECTOR

This time t_0 must be known in order to obtain exact values for the corrected V_R, in various identifications, and in the measurement of adsorption isotherms.

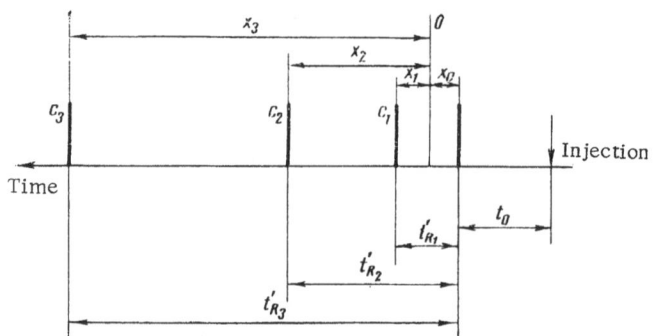

Fig. 185. Derivation of t_0 from the times of exit of three components of a homologous series when an ionization detector is used.

Various inorganic gases, e.g., the lighter inert gases, are used as being virtually unadsorbed; but ionization detectors do not record these, which therefore causes some difficulty.

Several indirect methods for t_0 have been proposed [69-74], which employ the linear relation of log t_R to the number of carbon atoms n_C in the molecule for the homologous series of normal hydrocarbons [75]. Then t_0 may [69, 70] be deduced from the t_R for three (or more) members of the homologous series.

The formula for the above relation for the n-alkanes is

$$n_C = b \log t'_R + g = b \log (t_R - t_0) + g,$$
(VI, 1)

in which t_R is measured from the time of injection, t'_R is the corrected time (measured from the instant of exit of the unadsorbed gas), and b and g are constants.

Peterson and Hirsch [69] used a method in which t_0 is calculated from the t_R for three members of a homologous series such that

$$n_{C_2} - n_{C_1} = n_{C_3} - n_{C_2}.$$
(VI, 2)

Figure 185 illustrates this method. The 0 line is drawn arbitrarily on the chromatogram, and the distances from this to the three peaks are measured (x_1, x_2, and x_3), x_0 being the distance to the point of exit of the unadsorbed gas. The t'_R are then $x_1 + x_0$, $x_2 + x_0$, $x_3 + x_0$.

Transformation of (VI, 1) gives

$$\frac{x_2 + x_0}{x_1 + x_0} = \frac{x_3 + x_0}{x_2 + x_0} \,, \qquad\qquad (VI, 3)$$

whence

$$x_0 = \frac{x_2^2 - x_1 x_3}{x_1 + x_3 - 2x_2} \,. \qquad\qquad (VI, 4)$$

From this x_0 we can calculate the t_R' for the n-alkanes. Gold's [70] lengthier graphical method may also be used with (VI, 1).

A direct method [74] uses methane as being a gas virtually unadsorbed at high column temperatures; but this method is not very exact, as methane is adsorbed appreciably by active materials.

Carrier gas saturated with organic vapor may be used to determine t_0 [76]; in this case the emergence of the unadsorbed gas produces a negative-going peak from the ionization detector (see also [78]).

LITERATURE CITED

1.   A. V. Kiselev, in collection: Gas Chromatography, Transactions of the First All-Union Conference on Gas Chromatography, Moscow, Izd. Akad. Nauk, SSSR, 1960, p. 45.
2.   A. V. Kiselev, Zh. Fiz. Khim., 35:233 (1961).
3.   A. V. Kiselev, Vestn. Mosk. Gos. Univ., Ser. Khim., No. 5, 1961, p. 29; No. 1, 1962, p. 3.
4.   A. V. Kiselev, in collection: Gas Chromatography 1964, A. Goldup (ed.), London, Institute of Petroleum, 1965, p. 238.
5.   L. D. Belyakova, A. V. Kiselev, and N. V. Kovaleva, Dokl. Akad. Nauk SSSR, 157:646 (1964).
6.   V. S. Vasil'eva, A.V. Kiselev, Yu. S. Nikitin, R. S. Petrova, and K. D. Shcherbakova, Zh. Fiz. Khim., 35:1386 (1961).
7.   A. V. Kiselev, I. L. Migunova, and Ya. I. Yashin, in collection: Gas Chromatography, Moscow, Izd. NIITÉKhIM, Issue 6, 1967, p. 84.
8.   C. G. Pope, Anal. Chem., 35:654 (1963).
9.   I. Halasz and C. Horvath, Anal. Chem., 36:2226 (1964).
10. I. Halasz and C. Horvath, Nature, 197:71 (1963).
11. D. D. Chegodaev, Z. K. Naumova, and U. S. Dunaevskaya, in collection: Fluoroplastics, Leningrad, Goskhimizdat, 1962.
12. D. Graham, J. Phys. Chem., 66:1815 (1962).
13. A. V. Kiselev and M. V. Serdobov, Kolloidn. Zh., 25:543 (1963).
14. O. L. Hollis, Anal. Chem., 38:309 (1966).

15. J. A. Favre and L. R. Kallenbach, Anal. Chem., 36:63 (1964).
16. D. A. Vyakhirev, A. I. Bruk, M. V. Zueva, and G. Ya. Mal'kova, Papers on Chemistry and Chemical Technology, Gor'kii, Issue 2, 1964, p. 268.
17. C. G. Scott, in collection: Gas Chromatography 1962, M. van Swaay (ed.), London, 1962, p. 46.
18. P. W.Solomon, Anal. Chem., 36:476 (1964).
19. L. B. Rogers and A. G. Altenau, Anal. Chem., 35:915 (1963).
20. A. G. Altenau and L. B. Rogers, Anal. Chem., 37:1432 (1965).
21. V. L. Keibal, A. V. Kiselev, I. M. Savinov, V. L. Khudyakov, K. D. Shcherbakova, and Ya. I. Yashin, Zh. Fiz. Khim., 41:2234 (1967).
22. N. N. Avgul', A. V. Kiselev, V. Ya. Mirskii, and M. V. Serdobov, Zh. Fiz. Khim., 42:1474 (1967).
23. A. V. Kiselev, in collection: Synthesis, Properties, and Uses of Zeolites, Moscow, Izd. Nauka, 1965, p. 13.
24. P. O. Sherwood, Erdöl Kohle, 16:843 (1963).
25. T. Nomura and M. Nukado, J. Chem. Soc. Japan, 64:810 (1961).
26. N. I. Lulova, L. I. Piguzova, A. I. Tarosov, and A. K. Fedosova, Khim i Tekhnol. Topliva i Masel, No. 5, 1962, p. 70.
27. T. G. Andronikashvili and Sh. D. Sabelashvili, in collection: Synthetic Zeolites, Moscow, Izd. Akad. Nauk SSSR, 1962, p. 65.
28. N. I. Lulova, L. I. Piguzova, A. I. Tarasov, and A. K. Fedosova, in collection: Synthetic Zeolites, Moscow, Izd. Akad. Nauk SSSR, 1962, p. 59.
29. K. J. Bombaugh, Nature, 197:1102 (1963).
30. J. E. Purcell, Nature, 201:1321 (1964).
31. R. D. Schwartz and D. J. Brasseaux, Anal. Chem., 29:1022 (1957).
32. I. N. Samsonova, S. P. Zhdanov, N. N. Buntar', E. V. Koromal'di, and V. A Golubeva, Zh. Prikl. Khim., 36:2502 (1963).
33. A. V. Kiselev, Yu. L. Chemen'kova, and Ya. I. Yashin, Neftekhimiya, 5: 418 (1965).
34. Z. Ya. Berestneva, T. A. Koretskaya, and V. A. Kargin, Kolloidn. Zh., 11: 369 (1949); 12:338 (1950).
35. K. D. Ashley and W. B. Innes, Ind. Eng. Chem., 44:2857 (1952).
36. A. B. Alexander and R. K. Iler, J. Phys. Chem., 57:932 (1953).
37. G. K. Boreskov, M. S. Borisova, V. A. Dzis'ko, O. M. Dzhigit, V. P. Dreving, A. V. Kiselev, and O. A. Likhacheva, Zh. Fiz. Khim., 22:603 (1948).
38. I. E. Neimark and R. Yu. Sheinfain, Kolloidn. Zh., 15:45, 145 (1953).
39. N. V. Akshinskaya, V. E. Beznogova, A. V. Kiselev, and Yu. S. Nikitin, Zh. Fiz. Khim., 36:2277 (1962); N. V. Akshinskaya, A. V. Kiselev, and Yu. S. Nikitin, Zh. Fiz. Khim., 37:927 (1963).
40. N. V. Akshinskaya, A. V. Kiselev, and Yu. S. Nikitin, Zh. Fiz. Khim., 38:488 (1964).
41. Z. Z. Vysotskii and I. E. Neimark, Ukr. Khim. Zh., 20:513 (1954).
42. K. V. Topchieva and G. M. Panchenkov, Uch. Zap. Mosk. Gos. Univ., No. 164, 1953, p. 13.
43. N. V. Akshinskaya, A. V. Kiselev, E. P. Kolendo, B. A. Lipkind, Yu. S. Nikitin, and Ya. I. Yashin, in collection: Gas Chromatography, Transac-

tions of the Third All-Union Conference on Gas Chromatography, Dzerzhinsk, Izd. Dzherzhinsk. Fil. OKBA, 1966, p. 222.

44. S. P. Zhdanov, A. V. Kiselev, and Ya. I. Yashin, in collection: Gas Chromatography, Moscow, Izd. NIITÉKhIM, Issue 4, 1964, p. 5.

45. D. P. Dobychin, N. V. Porshneva, and N. M. Turkel'taub, in collection: Gas Chromatography, Transactions of the Second All-Union Conference on Gas Chromatography, Moscow, Izd. Nauka, 1964, p. 69.

46. I. V. Grebenshchikov and O. S. Molchanova, Zh. Obshch. Khim., 12:588 (1942).

47. S. P. Zhdanov and E. V. Koromal'di, Izv. Akad. Nauk SSSR, Otd. Khim. Nauk, 1959, p. 811.

48. S. P. Zhdanov, Dokl. Akad. Nauk SSSR, 82:281 (1952).

49. S. P. Zhdanov, D. Sc. Thesis, Institute of Silicate Chemistry, Academy of Sciences of the USSR, Leningrad, 1959.

50. S. P. Zhdanov, A. V. Kiselev, and Ya. I. Yashin, Zh. Fiz. Khim., 37:1432 (1963).

51. S. P. Zhdanov, in collection: The Vitreous State, Transactions of the Third All-Union Conference, Leningrad, Izd. Akad. Nauk SSSR, 1960, p. 502.

52. S. P. Zhdanov, A. V. Kiselev, and Ya. I. Yashin, Neftekhimiya, 3:47 (1963).

53. M. M. Dubinin, Jubilee Symposium of the Academy of Sciences of the USSR, Vol. 1, Moscow, Izd. Akad. Nauk SSSR, 1947, p. 562.

54. M. M. Dubinin, Usp. Khim., 24:3 (1955).

55. D. H. Desty, J. H. Haresnape, and H. F. Whyman, Anal. Chem., 32:240 (1960).

56. V. I. Kalmanovskii, V. R. Lebedev, L. V. Polyakova, M. M. Fiks, and Ya. I. Yashin, Papers on Chemistry and Chemical Technology, Gor'kii, Issue 2, 1961, p. 351.

57. V. I. Kalmanovskii, A. V. Kiselev, V. R. Lebedev, I. M. Savinov, N. Ya. Smirnov, M. M. Fiks, and K. D. Shcherbakova, Zh. Fiz. Khim., 35:1386 (1961).

58. S. P. Zhdanov, V. I. Kalmanovskii, A. V. Kiselev, M. M. Fiks, and Ya. I. Yashin, Zh. Fiz. Khim., 36:1118 (1962).

59. M. Mohnke and W. Saffert, Gas Chromatography 1962, M. van Swaay (ed.), London, 1962, p. 214.

60. F. A. Bruner and G. P. Cartoni, Anal. Chem., 36:1522 (1964).

61. F. A. Bruner, G. P. Cartoni, and A. Liberti, Anal. Chem., 38:298 (1966).

62. F. A. Bruner and G. P. Cartoni, J. Chromatog., 18:390 (1965).

63. R. D. Schwartz, D. J. Brasseaux, and R. G. Mathews, Anal. Chem., 38:303 (1966).

64. R. D. Schwartz, D. J. Brasseaux, and G. R. Shoemake, Anal. Chem., 35:496 (1963).

65. J. Halasz and E. Heine, Anal. Chem., 37:495 (1965).

66. W. Schneider, H. Bruderreck, and I. Halasz, Anal. Chem., 36:1533 (1964).

67. F. Trusell and H. Diehl, Anal. Chem., 35:674 (1963).

68. A. J. de Rosset, Ind. Eng. Chem., 52:167 (1960).

69. M. L. Peterson and J. Hirsch, J. Lipid Res., 1:132 (1959).

70.   H. J. Gold, Anal. Chem., 34:132 (1962).
71.   M. B. Evans and J. F. Smith, J. Chromatog., 5:300 (1961).
72.   M. B. Evans and J. F. Smith, J. Chromatog., 6:293 (1961).
73.   J. F. Smith, Nature, 193:679 (1962).
74.   R. Kaiser, Gas Phase Chromatography, London, 1963, p. 51.
75.   A. J. M. Keulemans, Gas Chromatography [Russian translation], Moscow, IL,
       1959 [English edition: Second, New York, Reinhold, 1959].
76.   A. K. Hilmi, J. Chromatog., 17:407 (1965).
77.   N. K. Bebris, A. V. Kiselev, and Yu. S. Nikitin, Kolloidn. Zh., 29:326 (1967).
78.   R. Kaiser, Chromatographia, No. 5, p. 215 (1969).

Index